This Poisoned Earth

The Truth About Pesticides

NIGEL DUDLEY

PIATKUS

First published in 1987 by Judy Piatkus (Publishers) Ltd
5 Windmill Street, London W1

© 1987 Nigel Dudley

British Library Cataloguing in Publication Data
Dudley, Nigel
   This poisoned earth : the truth about
   pesticides.
   1. Pesticides——Environmental aspects
   I. Title
   632'.95042   TD196.P38

   ISBN 0-86188-632-1
   ISBN 0-86188-647-X Pbk

Designed by Paul Saunders
Phototypeset in 11/12 pt Linotron Sabon by
D. P. Media Limited, Hitchin
Printed & bound in Great Britain by
Mackays Ltd, Chatham

Contents

Acknowledgements

This book would not have been written without the help and support of a large number of people over the last few years. Although it is impossible to name them all, I would like to thank in particular Dave Buffin, Reg Green, Pete Hurst, Andrew Lees, Graham Martin, Dorothy Myers, Chris Rose, Pete Snell and Vic Thorpe for supplying information on various pesticides projects in the past, along with my colleagues at Earth Resources Research, the Soil Association and the Pesticides Trust. Other groups, especially in the Third World, have provided information through their own publications, often produced in far more difficult circumstances than we face in the North. Particular thanks to Audrey Hardy for providing me with a pleasant place to live in London through impecunious times. None of these people are responsible for any errors which may have crept into the text, or for the opinions, which are all my own.

Nigel Dudley, February 1987

Introduction

Every year, about a billion gallons of pesticide-containing liquid is sprayed on to farms, orchards, market gardens, parks, roadside verges and domestic gardens in Britain. Multiply that figure many, many times and you have an idea of the global pesticide consumption. Despite the fact that British crops are now treated with agrochemicals an average of six times each season (and sometimes literally dozens of applications are made) use is still increasing. In other parts of Europe, and in much of the Third World, pesticide use is just beginning the rapid growth experienced in Western Europe and the USA after the Second World War.

The range of chemicals used is bewildering. Hundreds of formulations, in countless mixtures, are sprayed as herbicides, fungicides, insecticides, acaricides (against spider and mites), rodenticides and molluscicides. In this book, 'pesticide' is used as a general term for all these categories. The Pesticide Safety Precaution Scheme in Britain every year lists literally thousands of brand name pesticides which have been cleared for use.

● The history of pesticides

Pests have been causing us problems in food production ever since we started to settle down in fixed communities and carry out farming. Of course, they have caused us more personal problems far longer than that! And efforts to control human parasites, and pests of crops and livestock, are older than history.

The modern era of pesticides built up in the four decades following the Second World War. The scientific revolution which caused this was the development of synthetic organic chemical pesticides, including many organochlorines, of which by far the most famous is DDT.

At this stage, DDT was seen as a 'miracle' chemical which could solve all farming problems at a swoop. It was thought to have low toxicity to humans (and was used against lice and other parasites) and a high toxicity to pests. Its long persistence in the environment, and action against a broad range of pests, was seen as a positive advantage, and it certainly appeared safer than many chemicals then being used, such as cyanide. And, to begin with, it worked, preventing many deaths from typhus and other diseases, and virtually eliminating malaria in some parts of the world through killing the mosquitoes which were carriers.

In the years since, pesticide use has expanded enormously, and it continues to grow to this day. Agricultural systems have developed which rely heavily on chemicals as an integral part of their functioning. Meanwhile, our attitudes towards pesticides have also changed, and some of the dangers in their use have been recognised. DDT, far from being a cure-all, has been exposed as extremely hazardous to wildlife, only dubiously effective in controlling pests and as having a number of hitherto unsuspected side effects on our own health. It has been banned in a number of countries, including Britain.

• The pesticides debate

Trade literature and official pronouncements place great emphasis on the good points about pesticides: the additional yields; the convenience; their role in modern, efficient farming. However, growing numbers of critics are pointing out negative aspects of pesticides as well. These principally revolve around the health and environmental dangers of pesticides, including risks to manufacturers, users and the public; the residues in food, and threats to the environment and other agricultural crops. But there are also more fundamental questions being asked about the overall efficacy of pesticides in improving agriculture and the degree of control of food production exercised by the huge transnational companies involved in the agrochemical industry.

Some of these issues were first exposed 25 years ago by Rachel Carson in her now classic book, *Silent Spring*. The barrage of controversy which this provoked resulted in several curbs on pesticide use. In the years since, it has often been assumed that pesticides had somehow been rendered 'safe' by these controls. Far from it. The two decades following *Silent Spring* saw a vast and unprecedented growth in the industry and, despite the efforts of a few tireless safety campaigners, the continued use and introduction of very hazardous chemicals. Even DDT, which Ms Carson identified as perhaps the most environmentally damaging of all agrochemicals, was only banned in Britain in October 1984. It is still used illegally today, and in 1985 destroyed a large heronry in Evesham, where birds had the highest levels of DDT in their bodies ever seen by Nature Conservancy Council pathologists.

Over the last few years, the debate about pesticide safety has been renewed in earnest. The successful suing of American chemical companies by Vietnam veterans for damage caused by the defoliant 'Agent Orange' (containing the pesticide 2,4,5-T) and the accident at the chemical plant at Bhopal have made worldwide headlines. In Britain, the Food and Environment Protection Act focused public and political attention on pesticides. Groups like the Soil Association, Friends of the Earth, the TGWU Pesticide Group, the London Food Commission and Oxfam have all produced major reports on certain aspects of pesticide hazards. Internationally, the Pesticides Action Network (PAN) provides a forum for groups interested in the issue. The South East Asian office for the International Organisation of Consumers Unions in Penang, Malaysia and the Environmental Liaison Centre in Nairobi, Kenya, provide important sources of information for workers in the Third World. The issues are being discussed more than ever before and a growing momentum of public pressure for change is developing.

The chemical industry has not taken the criticism lying down. They have adopted the time honoured methods of pushing the positive role of pesticides in hard sell advertising campaigns and rubbishing the opposition wherever possible. In 1986, a speaker at the 'Chemistry Industry After Bhopal' conference in the Hyde Park Hotel, London, likened the accident at Bhopal to the sinking of the *Titanic*, a one-off disaster which did not destroy the shipping industry. Repeated pronouncements link famine in Africa with the need for pesticides.

Elsewhere, industry spokespersons attempt to write off their critics. Mr. C. I. Smith, the forthright public relations officer of the National Association of Agricultural Contractors, describes pesticide critics (myself included) as 'beardy weirdies from the urban areas'. Making people appear 'different' in some way reduces their credibility. Elsewhere, industry accuses its detractors of amateurishness and bias. The methods of discrediting critics remain the same.

And, when all else fails, critics can always be accused of having radical or leftist subversive reasons for wanting the chemistry industry to 'fail' or 'collapse'. Poor Rachel Carson, a marine biologist already mortally ill when she wrote her book, was so accused in the 1960s (and chemical industry representatives also joked about her status as a 'spinster'). The same refrain is heard today. In 1984, Dr. Louis v Planta, chairman of the vast multinational Ciba-Geigy stated of pesticide critics that they were 'the common enemy' and that 'their declared objectives are often merely a front for aims that are fundamentally ideological'.

These tactics will no longer work today. Bhopal was not a 'one-off' disaster. The poisoning of the River Rhine produced more shock waves two years later, and public sensitivity means that far more accidents are reported than they were in the past. The people complaining in villages throughout Britain are not communists or anarchists speeding up the downfall of capitalism. They are ordinary men and women worried and angry about the risks to their health and to their childrens' health. This ill-feeling runs deep. At a recent Friends of the Earth workshop on pesticides, speakers reported getting anonymous phone calls from local people giving information about pesticide abuse once it was known they were running a campaign. The traditional reticence about speaking to outsiders was overcome by anger about what is happening. The increased flow of information is bringing growing pressure on the pesticide trade.

I am writing this book after spending several years working on pesticide issues for a number of environmental groups, principally the Soil Association and the newly founded Pesticides Trust. My personal experience being confined to pesticides in Britain, the emphasis will be on problems in the UK. But the international issues are so important that they are included in most chapters and there is a separate section on the problems connected with pesticides in the Third World.

The book has two aims. First, to outline the dangers that pesticides pose to humans, crops, livestock, wildlife and the environment. The text relies on current information, well-documented examples and published scientific data. And secondly, the book acts as a guide for anyone wanting to limit his or her exposure to the dangers of pesticides. Accordingly, at the end of each chapter is a section on recognising danger and/or fighting back against pesticide abuse, and a brief listing of actions which need to be taken on regional or national level to reduce pesticide risks. At the back I give details about books and addresses, and provide the first detailed checklist of pesticide hazards for the non-specialist. Detailed textual references are omitted, but the sources referred to under Books can provide those.

This book is polemical rather than a scientific paper. It brings together work which I have done myself, along with studies carried out by many other people throughout the world, working in academic institutions and non-governmental organisations. A general introduction of this kind cannot hope to be exhaustive or to act as more than a sampler to the issues. If I paint what might appear to be an overly black portrait of chemicals, it is to try to balance the overly optimistic view promoted by government and industry which we have all been living with for too long.

I have no doubt that pesticides will continue to be used for a long time yet. That is not the issue. But I believe wholeheartedly that this use should be tempered with far more precautions than at present, and with greater powers to protect public and users from hazardous chemicals. At the same time, and more importantly, a genuine change of heart is needed with regard to the way we use pesticides, so that minimising pesticide use becomes an essential part of the agricultural process. A major part of this must be the greater development of various organic farming methods that eliminate all but a small number of plant-based pesticides.

All the data mentioned in the text can be verified in the published literature, and the main sources are given in the book list. Details of chemical formula, and technical language, have been minimised to make this accessible to as wide a number of people as possible. It is hoped that this has not interfered with either the accuracy or the ease with which the book can be understood.

chapter one

Hazards of pesticides

In 1972, the World Health Organisation's Expert Committee on Insecticides estimated that there were approximately 500,000 cases of pesticide poisoning every year. About one per cent of these were thought to result in the victim's death, even in countries 'where medical treatments and antidotes are readily available'. It was considered very probable that higher mortality rates occurred in other countries, including most of the Third World.

The actual model used to calculate numbers of pesticide poisonings was constructed on a 'rather conservative basis', and estimated that there were 9,200 deaths per year. Oxfam researcher David Bull worked out that this meant some 6,700 deaths in the South. By 1980/81, recalculating these figures in the light of increased pesticide use, he suggested that there were 750,000 accidents per year, with 13,800 deaths, about 10,000 of them occurring in the Third World. Pesticide use is still increasing at an average rate of 12.5 per cent a year. This would mean that current (1986/87) annual accident levels, using the same premises, are running at well over a million people, and deaths approaching 20,000 a year. Given that much of the growth in pesticide use has been in the Third World – where medical facilities are poor, safety instructions minimal and chemicals 'dumped' after being banned in other countries – this figure is likely to be even more conservative than the original WHO estimate. (Since calculating these figures I have seen the same death toll quoted by Professor C. F. Wilkinson, of the Institute of Comparative and Environmental Toxicology in New York.)

In addition, longer term health effects, like cancer and birth abnormalities, are not usually included in statistics about pesticide damage. If these are also taken into account, the situation looks even more grim.

National governments and the international chemical industry like to claim that pesticides are basically safe. The calculations quoted above, based on original figures from the United Nations' World Health Organisation, suggest that well over 20 times as many people could be dying every year from pesticides as were killed at the accident in Bhopal. Yet because accidents are widely spread throughout the world, and often unrecognised as being caused by pesticides, they are largely ignored by doctors and politicians.

Pesticides can poison people in many different ways. Those who 'survive' may suffer permanent or recurring health problems. In this first chapter we look at the various types of hazards posed by pesticides, at some examples of hazardous agrochemicals widely used today, and at some case studies of people suffering from pesticide poisoning, both in Britain and other countries.

On 6th December 1986, the British government banned the sale of the pesticide **dinoseb**, and other related compounds, following a similar ban in the USA and pending a full re-examination. The decision was taken because of the pesticide's suspected role in causing birth defects. Donald Thompson MP, a Junior Agriculture Minister, claimed that the dinoseb-type chemicals were not dangerous to the consumer, but could be a hazard to people who applied spray commercially. He said 'we have taken this action particularly (because of) the possible effect on female operatives in the farming and horticulture industry'. Dinoseb is a highly poisonous chemical, which has been suspected of causing teratogenic effects (i.e. birth abnormalities) since at least 1973. It was previously cleared for use in agriculture in Britain, although it has apparently not been sold as a garden chemical in recent years. It had previously been withdrawn in Sweden in 1970 and in Norway in 1980.

There is nothing particularly special about the story of dinoseb. A newspaper report of its withdrawal caught my eye as I prepared this chapter, but I could have chosen from a fair number of alternative examples. Despite frequent claims that the pesticides we use are now more than acceptably safe, the steady trickle of products being banned or withdrawn shows that this 'safety' often has little basis in reality.

In 1985, the London Food Commission published a report on pesticides and food residues. The authors, Pete Snell and Kirsty Nicoll, combed the scientific literature to find evidence of long-term health hazards from British pesticides. Their survey, which they are the first to admit is incomplete, identifies 49 possible carcinogens (cancer-producing agents), 32 suspected teratogens (birth-defect producing agents), 61 suspected mutagens (capable of causing genetic damage) and 90 possible allergens or irritants (producing an allergic reaction or irritation to the skin, eyes and breathing). At the time, a government spokesperson dismissed the report as nonsense and claimed, once again, that no chemicals with long-term health effects are allowed on sale in Britain.

Since then, two pesticides have already been withdrawn in Britain for precisely those reasons (dinoseb, described above, and **ioxynil** which is discussed in the chapter on home and garden use). The clearance of a third, **aldicarb**, is urgently being reconsidered after new evidence suggested that it could damage or destroy the immune system. At least 38 of the pesticides used in Britain have been banned or severely restricted in one or more countries elsewhere. The British Government claims these bans are the result of over-caution, and that there are no real problems. Should we believe them? How many risks do you or I face when chemicals are used on farms or in gardens?

• *Pesticide poisons*

Pesticides are poisons. Their whole purpose is to kill insect pests, weeds, fungi and other unwanted plants and animals. Many of these chemicals are also directly poisonous to us, or to our pets and livestock. They may also have longer term effects on our health, such as those listed above, so that contact with pesticides today could do damage which does not manifest itself until many years in the future. Pinning down such long-term side effects to any one specific factor is never easy (and sometimes more than one factor can be involved anyway). Unfortunately, this allows hazardous chemicals to be used for a long time before their health risks are properly appreciated.

In the past, acute physical poisoning was the only pesticide hazard to be generally recognised. And it was recognised with good reason, because some extremely toxic chemicals used to be sold freely for application in farms and gardens: poisons like **mercury compounds, arsenic,** and **cyanide** have all been liberally sprinkled around the environment in an effort to kill pests at various times. Most have also finished off a few hapless farmers, gardeners and children in the process. Accidental poisoning was made more likely by the traditional practice of keeping pesticides in old pop bottles. Fortunately, this is now illegal in Britain (although it undoubtedly still sometimes occurs) and the number of fatal poisonings has decreased. This is not the case in the Third World, where direct poisoning is still a very serious problem.

Many of the pesticides which are most obviously poisonous have been banned or withdrawn in the rich countries of Europe and North America, and those remaining are more strictly controlled than in the past. However, this should not be taken to mean that there are no poisonous pesticides on the market. Far from it, many pesticides are directly harmful to humans in quite small amounts and some of the commonest are amongst the most poisonous.

> An American study has suggested that farmers exposed to herbicides for more than 20 days a year may run a sixfold increased risk of developing one particular form of cancer, Non-Hodgkins lymphoma. The herbicide to which farmers were most commonly exposed was **2,4-D** which is used in Britain. Researchers at the National Cancer Institute and the University of Kansas said that the results confirmed an earlier study from Sweden which also suggested a link between herbicide and this type of cancer.

• *The case of paraquat*

Paraquat is a herbicide manufactured by the UK-based Imperial Chemicals Industry (ICI), and at least 17 other companies, many based in Taiwan. It is one of the most commonly used herbicides in the world, and is especially popular in Britain. Over 270 tonnes of active ingredient were used annually in Britain between 1980 and 1983. It is frequently sold under the trade name of Gramoxone. Paraquat is a broad action herbicide, which will kill all vegetation and is used to clear new land for planting crops, as a general weedkiller and as a means of removing vegetation from paths and roadsides. Many garden chemicals are formulated from paraquat, including New Weedol and New Pathclear from ICI.

Paraquat is also a poison with no known antidote. If it is swallowed and gets into the lungs, paraquat begins an inexorable process of corrosion which is nearly always fatal unless treated within the first 24 hours. Death is slow, in the case of only a fairly small amount being swallowed frequently taking several weeks after swallowing the herbicide, and is usually intensely painful. The features of 'paraquat lung' can include 'honeycombing' of the lung and hardening of the breathing tract, until the lungs become almost solid with blood and mucus. Heart, kidneys and liver also sometimes suffer. As little as three to four mg can be fatal, and children are especially at risk.

There is also increasingly strong evidence that paraquat can damage health without even being swallowed. Doctors in Groote Schurr Hospital in Cape Town, South Africa, have described the death of vineyard workers from skin absorption of paraquat. Research in Denmark suggests that inhaling paraquat spray in an enclosed space can also cause problems. Scientists in Japan and Hungary have found high levels of paraquat in food grown on treated ground up to two years after application. A Russian research paper is reported to describe nose bleeding after inhaling paraquat fumes. The British 'Approved Products' handbook advises avoiding contact with eyes and skin. The British Crop Protection Council's *Pesticides Handbook* mentions that it is a suspected teratogen and mutagen. Although many authorities recommend wearing respirators and protective clothing while using paraquat, a survey of 43 Danish vineyard workers with 'paraquat lung' found no difference between those 'protected' and 'unprotected', suggesting that the respirators are of little use in practice.

> Proving damage is seldom easy. In June 1984, a mother and daughter in Wales were sprayed with herbicide being used as a dessicant against weeds on a neighbouring farm. Both suffered health problems, such as loss of energy, respiration and blood problems, and the younger woman developed a sore throat which lasted for about a year. The farmer initially admitted to using paraquat, but later changed his story and denied this. The incident is the subject of legal proceedings.

Because of its high toxicity, many countries have already banned or restricted the use of paraquat. It has been withdrawn in Denmark, confined to commercial users in New Zealand, restricted to institutional use in the Philippines, withdrawn from domestic use in Sweden, restricted in Turkey and banned in West Germany (because of fears of build up in the soil). Despite this response, paraquat is still sold in other countries all over the world. In its usual form it is a brown, colourless liquid which looks rather like cola. When stored in old soft drink bottles, it has been mistakenly drunk by children and adults. Stenching agents and emetics can be added, to dissuade would-be drinkers and make them vomit it straight up again if it is swallowed. However, these have apparently not always been used in the Third World.

No one denies that paraquat is virtually always fatal if ingested directly. There is a continuing debate about the toxicity of paraquat if breathed in from spray drift or used in confined spaces. ICI's own research concluded that 'if a man spends much of his time walking through spray drift prevented from dispersal by overhanging foliage there may be a risk of lung damage'. Yet paraquat is still available over the counter of any garden supply shop in Britain.

• Defining pesticide poisons

I have dwelt on paraquat at some length because of its popularity and the controversy which has long surrounded its use.

But paraquat is by no means the only poisonous chemical on the market and the safety record of ICI is relatively high compared to other chemical companies.

The commonest measure of toxicity is the LD50 test. The LD50 is defined as the weight of active ingredient needed to kill 50 per cent of the test animals exposed to the pesticide. This rather macabre measuring stick is usually defined in units of milligrams of pesticide per kilogram body weight and commonly involves the use of mice or rats. It is not a wholly accurate assessment of the effects on humans, certainly does not identify hazardous chemicals very clearly, and there are often several LD50s for any particular chemical. However, anything with an LD50 of less than 100 mg/kg can be regarded as highly toxic to humans. The London Food Commission found 70 British pesticides in this category, including **azinphos methyl** (13 mg/kg), **captan** (10 mg/kg), **chlorfenvinphos** (9.9 mg/kg), **diquat** (30 mg/kg), **methiocarb** (40 mg/kg) and **warfarin** (1 mg/kg), as tested on mice and rats.

Effects on health from these and other pesticides are numerous. They can include neurological damage, liver and kidney effects, changes to the blood plasma, internal bleeding and the development of allergies. A recent British trade union study concluded that 'there are major and widespread health and safety problems connected with the use of pesticides'.

On 14th July 1977, farmer Enfys Chapman was sprayed by a helicopter treating a pea field bordering her smallholding near Ely, in Cambridgeshire. The pilot was using Hostathion, a trade name for the pesticide **triazophos** marketed by Hoechst. Mrs Chapman was covered in spray herself, and probably further contaminated while she treated her rapidly sickening cattle over the next five days.

A number of cattle died quite quickly, from a combination of bloat (where they lost all strength in their back legs), mineral deficiencies induced by poisoning (especially selenium) and fatty livers. MAFF autopsies also showed bone deterioration. Some 'surviving' cattle became sterile and had to be culled.

Five days after the incident, Mrs. Chapman became so ill herself that she was admitted to Addenbroke's

Hospital. When I met her, in 1984, she was still virtually confined to a wheelchair, and had been warned by the Health and Safety Executive to move to an urban area so as to avoid spray drift, because the poisoning had made her sensitive to pesticides. She had been forced to move into Cambridge. She suffered spasms if affected by drift, and one of these attacks had resulted in a coronary in 1983.

Mrs. Chapman had also suffered periods of delirium, in which she forgot the English language and reverted to her native Welsh. Her memory was dulled, she had difficulty in concentrating and described herself as a 'victim of premature ageing'. For years she had muscular cramps several times a night, sometimes lying rigid for up to an hour or being thrown completely out of bed.

Hostathion is still marketed by Hoechst, who have never admitted liability for Enfys Chapman's condition. They did pay her £12,000, however, in an out of court settlement.

● *Irritations and allergies*

Have you ever suddenly come down with a bout of hay fever after working in the garden or walking through farmland, even if you are not usually a sufferer? Hay fever is traditionally caused by pollen irritating the lungs, hence its name. Nowadays, identical symptoms can be an indication of allergic responses to a large number of artificial chemicals as well, including many pesticides.

Flicking through the British government's now-defunct guide to pesticide use and safety – the *Approved Products for Farmers and Growers* – quickly shows that the commonest caution listed against agrochemical products reads something like, 'can be irritating to skin, eyes and respiration'. This innocuous phrase means that coming into contact with the pesticide can cause streaming eyes, burning throat and, if it splashes on to the skin, some form of dermal irritation. Something rather like a cold in fact. Sub-acute, irritating effects like this are probably the commonest results of pesticide contamination encountered in Britain today. Yet they are also probably the least recognised. How many doctors would suggest that you were suffering from chemical contamination if you visited them with a summer cold? And how many people would bother to consult a doctor in the first place?

The degree of irritation obviously differs between chemicals. Some people are far more sensitive than others, so that not everyone will suffer the same effects, or react the same way each time they are contaminated. But literally hundreds of products sold today carrying warnings about irritation. And irritation of this kind (or more serious) was reported by half the farm workers interviewed in a recent trade union survey of pesticide use, despite the fact that they are supposed to be provided with special protective clothing and to know exactly how to handle the chemicals they apply.

• *Chronic effects*

Unfortunately, it is not just relatively minor ailments that crop up, frequently unsuspected, amongst people coming into contact with pesticides. Far more serious is the possibility that pesticides can also have longer term effects on our own health, and on the health of future generations. Three 'chronic effects' are usually included in this category: carcinogens, teratogens and mutagens.

Cancer

Carcinogens are substances which can produce cancer. Famous examples of carcinogens are some kinds of radiation (including gamma radiation and sunlight); various components of cigarettes and asbestos dust. There are thousands of others, including naturally occurring carcinogens like those found in certain kinds of rock, in the mould which grows on peanuts and other plants and in bracken. However, today these natural carcinogens are far outnumbered by the large number of artificial carcinogens produced accidentally as a by-product of industrial society, including industrial chemical by-products, pollutants, food additives and pesticides amongst many others.

Proving that a substance is carcinogenic is a lengthy and difficult process, as shown by the decades it took to 'prove' the link between smoking and cancer (a link which the tobacco industry still denies against all the evidence). Tests on laboratory animals are inconclusive and epidemiological studies on human populations take a great deal of time and money. The latter can also only be used after a substance has been in the environment for decades, so that in assessing the risks of a new industrial chemical, such as a pesticide, the laboratory tests are the best method currently available.

We know that some pesticides in use today are carcinogens. There are strong possibilities that others exist. The 50 or so suspected carcinogens identified by the London Food Commission are unlikely to all be cancer-producing in humans, of course. Some will be so weakly carcinogenic as to be of little hazard to ourselves. (There is no safe level for carcinogens, and all safety decisions have to be made on balance of probabilities.) On the other hand, it is very probable that other pesticides are carcinogenic but have simply not been studied in enough detail as yet to identify them. Many of the pesticides currently in use today are thought unlikely to pass the more rigorous safety testing now legally necessary for new products; it was this kind of 'retesting' that identified hazards in ioxynil and dinoseb as mentioned earlier.

Take the fungicide **captan** for instance. Captan is a spray used on farms, orchards and private gardens to control black spot on rose, scab on apple and pear, and stem rot on tomato. British products containing captan include Murphy's Captan 83, Murphy's Hormone Rooting Powder, May and Baker's Strike Rooting Powder, etc. Captan has been identified as a potential carcinogen by the IARC and NIOSH. Carcinogenic risks were listed as reasons for banning its use in domestic gardens in Sweden, for its complete banning in Finland, and for withdrawing its registration in March 1986 in West Germany. Norway has already severely restricted sales. Russian studies showed evidence of genetic risks. Yet captan still remains openly on sale in Britain, where it is used in farms and gardens.

Cancer experts disagree about the importance of various factors in producing cancer in humans. Richard Doll and Richard Peto, of the Imperial Cancer Research Institute, argue that most cancers are related to smoking (30 per cent) and diet (35 per cent) with additional factors being alcohol, food additives, sunshine and 'sexual habits', with industrial pollution being relatively unimportant. Samuel Epstein, Professor of Occupational and Environmental Medicine at the University of Illinois, puts the occupational component of cancer at a minimum of 20-40 per cent, a view supported by a group of American Federal experts. These two theories are known as the 'lifestyle' theory and the 'work related' theory.

Proving conclusively how much pesticides contribute to cancer levels is clearly impossible given this amount of disagreement. Many doctors now believe that development of cancer is a multifactorial phenomenon, so that several different carcinogens can be involved in any one cancer. Most medical experts agree that minimising contact with carcinogens is an essential first step towards reducing risk. The carcinogens in pesticides are one of the risk elements which could be most easily reduced or eliminated.

Our children at risk

The other long-term effects centre on future populations. Mutagenic substances can cause mutations in the genetic make-up of the cell, which can appear as weaknesses in future generations. Teratogenic chemicals increase the risk of birth defects developing in the children of people who have been contaminated. Chemicals suspected of causing genetic effects in Britain include **amintriazole** (also a carcinogen), **captan**, **chlordane** (which is suspected of causing all three long-term effects), **dinoseb** and **fenitrothion** (both suspected mutagens), **malathion**, **simazine** and **thiram**. There are many more.

> **Diazinon** is an organophosphorus insecticide and acaricide (i.e. spider and mite poison) used especially against cabbage root fly, carrot fly and mushroom fly. It was listed in the British government's *Approved Products* book and classified as only 'moderately hazardous' by the World Health Organisation. British

trade names include Basudin 40 WP from Ciba-Geigy and Murphy's Root Guard (a garden chemical). However, it has been mentioned as a possible tera- togen by several sources, including NIOSH in the USA which recommends care when being handled by women of childbearing age. Hungarian studies suggest that there is an increase in chromosomal structural deformations in exposed workers. The International Labour Organisation (another United Nations body) lists it as having embryotoxic effects and classifies it as being highly toxic. Yet it is still widely available and is on record as even being used inappropriately on house plants, where it caused poisoning.

Dinoseb and ioxynil have both been withdrawn from some uses in Britain recently because of suspected birth effects although the ban on dinoseb was later, temporarily, reversed. But other similar chemicals remain on the market.

Allergy and disease

No one doubts that pesticides can cause serious health prob- lems, although as we have seen there is a lot of room for debate about the levels of risk and which chemicals are dangerous. Today, a much more controversial theory is being discussed amongst a minority of doctors, suggesting that pesticides (along with other chemicals) are causal factors in a complex relationship between diet, allergy and disease, which can cause a fairly complete breakdown of health in some cases.

Dr. Jean Monro, a specialist at London's Nightingale Hos- pital, is the leading British supporter of this theory, which is known as 'clinical ecology'. She claims that about 15 per cent of the population are likely to suffer badly from allergies and that a further 15 per cent can be 'sensitised' to a range of allergies through contact with allergenic chemicals, including many pesticides. One possible reason suggested is that they lack enzymes needed to 'detoxify' the potentially harmful chemicals which are ingested in everyday life in the modern world. Dr Monro believes that the rising numbers of extreme allergies she sees in her professional work stem largely from the increase in allergenic chemicals. She recommends a process of 'detoxification' as the only way of dealing with this. Her patients are advised to go to clean environments (perhaps in the mountains), eat an extremely careful diet, with totally organic food and containing as little repetition of particular types of food as possible. She also recommends avoiding any contact with plastics or other artificial materials, synthetic fabrics and so on. After a period of detoxification, the patient may have a chance of surviving, with care, in the everyday, polluted world again.

Dr. Monro's closest professional colleague in the USA is Dr. Bill Rea, who is based in Dallas, Texas. He has taken the detoxification principle a step further, by building a clinic where patients can recuperate in totally pristine conditions. He claims very high success rates in dealing with sufferers of extreme allergy problems.

Whether the theory behind this treatment is true or not has yet to be proved. Sceptical doctors point out that the problems treated by clinical ecologists could be psychosomatic. But the effects of low level contamination of this sort, by pesticides or almost any other chemical pollutants, remain almost completely unexplored. We will return to this issue in the next chapter, on spraying problems.

Pesticide hazards in Britain have recently been under review by the House of Commons Select Committee on Agriculture, which, interestingly, extended its normal six month time limit for evidence because of the wide range of issues which came up. Sir Richard Body, the Conservative chairman of the Committee, said 'it would be totally wrong to have a superficial enquiry. The more we have looked into some of the criticisms of pesticides and government regulations the more we realise that we are engaged in an important enquiry for agriculture, the environment and the nation's health.'

● *Pesticide hazards in the Third World*

'A number of people get affected by occupational pesticide poisoning, but the people don't usually report to the hospital until it's too late. Also, the sufferer is usually also suffering from other health problems which contribute to death. This means that one of the

> other causes will usually be recorded as the cause of death. . . Pesticides are often mixed by hand and sometimes pesticide powders are just sprinkled by hand on to the crops.' Mr. S. Selliah, Asian representative of the International Federation of Allied and Agricultural Workers, interviewed by David Bull, formerly of Oxfam.

Any health problems we face from pesticides in the North are multiplied many times over in the South. At a time when a lot of people in the rich countries are actively trying to help in the vital process of increasing the availability of food in the poor countries, the damage done by poorly designed agricultural systems, including misuse of pesticides, is a scandal of global proportions. I will be returning to this many times in the course of the book. For the moment, it is important to stress how many more accidents are likely to occur in the Third World countries than in our own. When the World Health Organisation made its estimate of pesticide accidents which I quoted at the start of this chapter, the 500,000 injuries per year had a margin of error from 250,000 to 1,453,000; i.e. there might only be half as many cases, but could be up to three times as many within their prediction model. And this is regarded as a conservative estimate.

> Figures for pesticide poisoning are notoriously difficult to come by in the Third World, and researchers have to extrapolate from the few examples that get adequately researched and written up. In 1978, 15,504 people were admitted to government hospitals in Sri Lanka with pesticide poisoning, and 1,029 of them subsequently died. Tragically, 70 per cent of these fatal poisonings were thought to be the result of suicide bids. However, this still leaves over 300 accidental deaths. It must also be realised that most minor cases of pesticide poisoning never get to hospital, and that long-term effects are ignored in these statistics. In 1977, pesticides are thought to have killed almost twice as many people in Sri Lanka as malaria, tetanus, diphtheria, whooping cough and poliomyelitis combined.

Problems occur in many other countries as well, although few details filter through to the media or politicians. In the Pacific coastal regions of Central America, more than 14,000 pesticide poisonings were tabulated between 1972 and 1975. In Pakistan, at least five people died, and almost three thousand became ill as a result of **malathion** used to control malaria in 1976. In Guatemala, many of the thousands of itinerant farm workers who harvest coffee, sugar and cotton become ill every year from organophosphate poisoning and one nurse claimed that her hospital treated 30 to 40 pesticide victims a day at the height of the season. In El Salvador, a study found that there were six times as many deaths from pesticides as the data collected from hospitals suggested. A doctor in the Philippines treated four to six patients a day for pesticide poisoning in the peak growing season. Poisoning effects may be worse in people suffering from malnutrition.

Dr. Y. R. Reddy of the Niloufa Children's Hospital in Hyderabad, India, described the reasons for high rates of pesticide accidents:

'. . . poor people live in one room only, so they keep their pesticides in the room as well. They live there, cook there and eat there. The powder is in the air when they spray their crops, the spray sometimes drifts into the house, and children are exposed. . .'

Dr. Reddy has pinpointed one of the main reasons for the vast toll of pesticide poisonings in the Third World; safeguards are usually appallingly inadequate. A large part of this problem is due to lack of any information, or education, for spray users. Pesticide containers are unlabelled, or warning instructions printed in a foreign language like English or French. Even if safety messages are written in the native tongue, many users will be unable to read. Pesticides are often mixed by hand, and hands are used for rubbing eyes, eating food or cupped to drink from. Protective clothing is unavailable or it is too hot to wear heavy clothes; this can cause heat exhaustion in tropical climates. Governments are often unable, or unwilling, to enact proper safety legislation.

In the Philippines, farmers are advised to wear protective clothing when they apply pesticides. They usually don't. A saleswoman from Planter's Products Inc. told the Farmer's Assistance Board of the Philippines that 'farmers are stubborn people. They do not listen to our advice.' However, when asked whether her store sold protective equipment, she admitted that it did not and that 'they won't buy anything since they don't have the money'.

However, there are other, clearly avoidable, problems which relate far more closely to ourselves. Pesticide sales representatives often push their products very unscrupulously in Third World countries. Hazardous chemicals, banned in the North, continue to be sold in the South, often by the same companies. Third World governments are kept in ignorance of the hazards. A system of agriculture is pushed by aid agencies and 'developers' which relies on heavy pesticide use. These issues will be returned to again in the chapter on the Third World.

The Dirty Dozen

The Pesticide Action Network (PAN) is a worldwide coalition of groups and individuals who are opposed to the irrational spread and misuse of poisonous pesticides. It has regional offices in North America (USA), Africa (Kenya), Latin America (Ecuador), Asia/Pacific (Malaysia) and Europe (Brussels). On 5th June 1985, World Environment Day, PAN launched an international 'Dirty Dozen' campaign against hazardous chemicals. This was the first global campaign for the elimination of certain pesticides judged to be too dangerous to justify using for crop protection.

Anwar Fazel, Director of the International Organisation of Consumers' Unions (IOCU) regional offices in Penang, Malaysia, explained the basis of the campaign: 'We selected these twelve pesticides not only for their toxicity, but for the unique hazards they pose to people in the Third World countries. The problem, however, is global. These chemicals are mainly used on 'export crops' – such as bananas, coffee, tea, cotton and rubber – and therefore return to consumers in Europe, the USA and Japan as residues in food and commodities in a "circle of poison".'

The 'Dirty Dozen' campaign had four main aims:

1. to ensure that human safety and environmental health are considered foremost in all policy decisions affecting pesticide use and trade.

2. to end the use of the 'Dirty Dozen' wherever their safe use cannot be ensured.

3. to eliminate double standards in the global pesticide trade by obtaining open public access to technical data, especially health and safety information and facilitating the advent of effective controls wherever pesticides are manufactured and/or used.

4. to generate public and institutional support for research into alternative pest control programmes, and for their use in practice. These programmes should minimise pesticide use. Many such programmes already exist.

The 'Dirty Dozen' pesticides selected were: camphechlor; chlordane/heptachlor; chlordimeform; dibromochloropropane or DBCP; DDT; the 'drins', aldrin, dieldrin and endrin; ethylene dibromide or EDB; HCH/lindane; paraquat; ethyl parathion or parathion; pentachlorophenol or PCP; and 2,4,5-T.

The 'Dirty Dozen' campaign involves groups in at least 50 countries, with a combined membership running into many millions of people. It is probably the most concerted attack yet on the sale and use of unacceptably hazardous chemicals.

● *Clearance of pesticides in Britain*

In the last couple of years, there have been efforts to clear up some of the confusion surrounding clearance of pesticides in Britain. Some of the established practices have been revised and one particular publication abandoned.

Today, all pesticides used must be passed under the Pesticides Safety Precaution Scheme, which has existed for years as a voluntary scheme but now enjoys full legal status. The scheme is run by the Ministry of Agriculture, Fisheries and Food (MAFF) and administered by a team of experts known as the Advisory Committee on Pesticides (ACP). The ACP is, in turn, advised by a series of subgroups (which is currently expanding), the most important being the Scientific Sub-Committee (SSC).

Although the ACP and SSC are supposed to be entirely separate from the industry, and members have to make a declaration that they have no commercial interests which would undermine their objectivity, the government has re-fused on a number of occasions to state categorically that no links between members and the chemical industry ever exist. Given the close links between industry and academia, and the relatively small world of pesticide research, interaction be-tween the two is inevitable, even if no money is involved. The role of secrecy in the ACP is discussed in the chapter on politics.

> Advertising is one of the more contentious issues sur-rounding the health hazards of pesticides. Calls for more information about hazards, including long-term health hazards have always been rebuffed by succes-sive governments, but pressure for this to be included is growing all the time. The Earl Peel, a Conservative peer, said during the 1985 Food and Environment Protection Bill that: 'If we can have stated on a packet of cigarettes that smoking can dangerously affect your health, there is no reason why we should not have a warning on a fungicide that it can seriously affect the health of insects. . .'
>
> In practice, not only are chronic health effects ignored, but advertisements often promote an un-realistically safe picture of pesticide use. While prepar-ing evidence for the Soil Association's report on gar-den chemicals, I found several advertisements where sprayers were being used in an obviously unsafe fashion. In one, the woman spraying was holding the sprayer up in a fashion which would have ensured she get a face full of chemical if there had been any wind. Others show children playing in a garden which has been freshly sprayed.
>
> Over the last 30 years, advertising in farming jour-nals has also increased enormously, and adverts play

up the 'macho' image of pesticides rather than safety or environment. Chemicals called Assassin, Clout and Sting are depicted in very aggressive images; one popular one includes a gun in a holster. Yet these advertisements often do not even mention the active ingredient contained in the particular pesticide, so farmers have little opportunity to reject any which they consider are unsafe.

● *But is it really a problem?*

So what? If you eat half a pound of table salt at a sitting you stand a good chance of dying, as the chemical companies are quick to tell their critics. Anything tested rigorously enough, they argue, can be shown to have carcinogenic effects. If pesticides are such a problem, why don't we see the bodies piling up?

It is true that there are relatively few cases of immediate death from pesticide poisoning in the more technically advanced countries. In part this is due to public health campaigns of past years, to greater literacy and education amongst users and the banning of the most obviously toxic substances. But that is far from saying that pesticides are 'safe'. As we have seen, the number of chemicals suspected of causing long-term effects are very considerable and these chronic effects have been inadequately studied. The enormous carnage wreaked by pesticides in many Third World countries is no longer open to serious doubt. The evidence for their role in long-term deaths in the richer nations is growing all the time. There is an unarguable case for minimising our contact with these potentially deadly chemicals as far as humanly possible. That this ideal is never anywhere near realised is the subject of much of the rest of this book.

The potential for hazardous chemicals to 'slip through the net' of safety controls is emphasised by the number which have already been withdrawn in Britain. In an answer to a parliamentary question by Labour MP Dale Campbell Savours, in February 1985, the government listed 20 kinds of pesticides whose safety clearance had been withdrawn under the Pesticides Safety Precaution Scheme, for safety reasons, since the inception of the scheme. The 20 were: **antu, azobenzene, cadmium compounds, calcium arsenate, chlordecone, DDT, endrin, ethylene dibromide, hexachlorobenzene, inorganic fluorides, mercuric chloride, methyl mercury, nitrofen, phenylmercury salioylate, potassium arsenate, selenium compounds, sodium arsenite and 1,1 2,2-tertrachloroethane.**

Since then, at least two more, **dinoseb** and **ioxynil**, have been more carefully controlled and the use of **dieldrin** for wood treatment is no longer cleared. Meanwhile, withdrawals of chemicals used in Britain continue in other countries. In 1985, the pesticides **captafol, captan** and **folpet** were withdrawn in West Germany although all three are still available in Britain.

• *What you can do*

This chapter has simply outlined some of the problems concerning pesticides. Later chapters look more specifically at different ways in which you can come into contact with these.

The first stage in starting to take pesticides more seriously is to identify which ones are the most hazardous; i.e. becoming more aware of the dangers involved. However, recognising pesticide hazards is by no means easy. Information is frequently inadequate, labels confused and confusing and details about long-term health effects are omitted altogether. (Government officials in Britain stick to the line that this is because no chemicals with long-term health effects are ever cleared for use here, but as we have seen this is a very unrealistically optimistic claim.)

If you are going to use pesticides, you would be well advised to do a little detective work on the possible hazards before you start. We try to give you a head start on this process by giving a summary table in the Appendix of some of the hazards associated with over 200 pesticides, but this is inevitably superficial and incomplete. This kind of information needs to be backed up by a certain amount of knowledge about how to interpret the data that you are actually presented with. The following table contains a guide to what you might see on labels, or in books, relating to specific chemicals, and how these might be interpreted.

As a general rule, remain suspicious! Labels are so cramped that it is easy to miss things. The World Resources Institute found that, for the USA, different products containing the same active ingredients have been found to have contradictory first aid advice on their labels. Comparing British and American sources of information on hazards during the preparation of this book has shown great differences in opinion about what chemicals pose what risks. Even official British sources do not always appear to agree. And, as I have said before, not all the dangers will have been spotted in the first place.

The following symbols are international hazard symbols which also give valuable advice (although why two of these symbols are identical is rather a mystery!). Look out for them on pesticide packets and bottles.

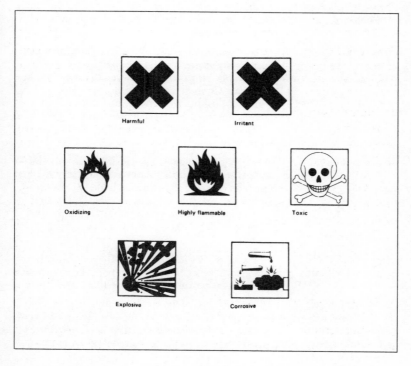

How to interpret information about pesticide hazards

Warning	What it means
'Chemicals included in the *Poisonous Substances in Agriculture Regulations*'	These chemicals are classified as particularly poisonous and users have to abide by certain precautions outlined in *A Guide to the Poisonous Substances in Agriculture Regulations 1984* available from HMSO. Substances are classified from Part I (most poisonous) to Part III and regulations include types of clothing, etc. Should not usually be available for home use, but a few are in practice.
'Chemicals subject to the Poison Rules'	Subject to the provisions in the Poisons Act of 1972. The *Poison Rules* are also obtainable from HMSO and include general or specific provisions for the labelling, storage and sale of scheduled poisons. (Several garden chemicals are included here, such as paraquat and tar oil.) Any pesticides labelled under the *Poison Rules* will be highly toxic.
'Anti-cholinesterase compounds'	Includes organophosphorus compounds and some carbamate pesticides. Particularly poisonous to susceptible people.
'Irritating to eyes, skin and/or respiration'	Very variable response likely depending on how sensitive you are. Should take particular care to avoid splashing on skin and eyes, or being caught in spray. Contamination should be treated by 15 minutes flushing with cold water and medical treatment if symptoms of irritation develop. (These can sometimes appear like a cold and sore throat.)
'Harvest interval'	Usually given in days. The gap between spraying and when a crop can be harvested and sold. This is a useful clue that the chemical is likely to be both toxic and persistent. It is especially significant if pesticides with a harvest interval drift onto other crops which may be eaten during the danger period.
'Dangerous to fish, game and wildlife'	Or something similar. If there are warnings about danger to larger birds or mammals, the chances are that it won't do us much good either.
'Minimal interval to be observed between application and access to treated areas'	Listed in days for livestock, poultry and/or humans. This means that the pesticide is toxic and, probably, that it has toxic vapour as well. (Often used for pesticides used in greenhouses, etc.)

Warning	What it means
'Do not use treated seed for human or animal consumption'	Again this means that the pesticide is toxic (although it tends to be listed for all treated seeds).
Chronic effects	Never listed on labels. But we do, check the Appendix.

● What the government could do

This chapter has outlined some of the more worrying aspects of pesticide hazards. It has illustrated gaps in legislation, and a failure to address many issues with the attention they deserve. It has also shown that there are clear needs for changes in legislation with respect to some aspects of pesticide safety. Some of the most important of these are listed below:

● a ban on the more hazardous chemicals currently on the market. As of 1st January 1985, there were well over 500 pesticide products cleared in Britain which contained one or more chemicals which had been banned or severely restricted in a number of other countries. Britain is fast acquiring a reputation for having lax regulations with regard to pesticides, and further bans are urgently required. These include, amongst others, aldrin, chlordane, lindane, paraquat and 2,4,5-T.

● it should be an offence to sell pesticides for an unregistered use. This now looks much more likely to become law with the legal obligation for all pesticides to be cleared under the Pesticides Safety Precaution Scheme (previously voluntary), but still leaves worrying gaps with respect to pesticides used for minor crops, or used under test conditions.

● labelling of pesticides should include extra details about hazards. In particular, suspected long-term effects should be identified clearly, in the same way as they are on cigarette packets.

● legal controls on pesticide advertising should be strengthened to include a ban on advertisements which show

pesticides being used in an unsafe manner, and an obligation to state ingredients, and put hazard warnings on all
advertisements.

- there should be public access to the information used in
  deciding whether a pesticide is safe by the Advisory Committee on Pesticides. The experience in the USA has shown
  that this does not give rise to serious problems of professional trade secrets being lost in practice.

- the Advisory Committee on Pesticides should evaluate test
  data from other countries where applicable. This would
  greatly increase the amount of safety data available for
  decision making.

- systematic epidemiological surveys of the health of farmers,
  agricultural workers, growers and people manufacturing
  pesticides are needed. The large body of anecdotal evidence
  and growing number of medical studies show that pesticide
  effects on health are real problems; these will never be fully
  appreciated until proper, scientifically valid and sufficiently
  large surveys are carried out. This will only occur with
  proper government funding.

chapter two

Problems in using pesticides

During a hot spell in June 1984, a crop-spraying plane overshot its intended target and released a fine yellow mist of fungicide above the village of Blackhall, near Durham. Scores of people were covered with pesticide. Fifteen were affected so badly that they needed hospital treatment for a range of side effects, including headaches, skin rashes, sore throats and eye infections. The fungicide was of a type chemically similar to certain nerve gases, and villagers were advised not to eat surface-growing garden vegetables for at least a month. Environmental health officers said the chemical could have particularly bad effects on asthmatics and the elderly.

One villager complained: 'The low flying has gone on for years. The plane flew over here so low that I picked up my two children and ran out of the house because I thought he was going to crash into it.' Yet the Civil Aviation Authority refused to prosecute the pilot, and contented themselves with a written warning, which is kept on their own records but is not even noted on the pilot's licence. The Health and Safety Executive, in an independent investigation, concluded that the incident was due to pilot error. The CAA defended their decision to do nothing by saying that the pilot's actions were not intentional, and they declined even to release his name. 'As far as we are concerned', they stated, 'the matter is now closed.'

It is not closed, of course. It lives on in the minds of villagers who were ill, or who lost their carefully nurtured vegetables. By chance, the accident occurred on the day after the Soil Association published a report on spray drift, *Pall of Poison*, and local television and radio stations gave publicity to both events simultaneously. Newspapers carried the story and its aftermath. Residents' anger at their treatment, and the lack of any punishment for the pilot, showed through in letters and articles. A local incident became one more case study in the growing evidence about the harmful effects which occur when pesticides are used indiscriminately on the land.

> The problems of spraying in the small Hertfordshire village of Kimpton became so bad that in the summer of 1985 almost a hundred people signed a petition saying, AERIAL CROP SPRAYING CAN DAMAGE YOUR HEALTH BUT CARRIES NO GOVERN-MENT HEALTH WARNING! People complained to Friends of the Earth about being sprayed directly by aircraft while walking along public roads, not being informed about aerial spraying, or being given the wrong day of application. Several people claimed to have suffered increased asthma problems and colds as a result. The spray contractors said, 'we have great faith in the vetting procedures for the chemicals, and have never yet poisoned any animal or person'.

We have shown in the previous chapter that pesticides are hazardous. It follows that careless or indiscriminate use of pesticides will almost inevitably result in danger to the health of people, crops, livestock and wildlife. Many of these risks have never been properly quantified. In this chapter, some of the dangers of using pesticides are assessed. A number of well documented case studies show that this is not merely a theoretical problem, but one which is facing people right now, and on a much wider scale than official figures suggest. The chapter concludes with a detailed guide to how you can minimise your own risk of being contaminated by pesticides. And it gives advice about how to fight back against the dangers of illegal spraying or contaminated food if you do suffer problems.

The main issues discussed here are: direct health risks

among users; drift of pesticide spray; the specific case of aerial spraying; and, last but by no means least, the risks from chronic exposure to pesticides in the environment. None of these issues are new. The dangers posed by pesticides have been the cause of rumbling discontent in rural areas for many years. Today, for the first time, they are starting to be brought out into the open and discussed by doctors, ecologists and politicians.

● *Dangers for pesticide users*

Chemical apologists argue that official statistics showing a low incidence of pesticide illness among farm workers 'proves' that agrochemicals do not pose appreciable risks to health. The logic of this complacent viewpoint is flawed on two counts. First, a properly trained and responsible pesticide operator, wearing protective clothing, should in theory be less at risk than a member of the public caught unprotected in spray drift, or someone who comes into contact with contaminated vegetables. (In fact, users often take minimal precautions, as we shall see below.) And secondly, there is now ample evidence that the incidence of damage to users is far more frequent than official figures suggest anyway.

Dr. Joyce Tait, a senior researcher at the Open University, conducted a survey of 80 farmers who regularly used pesticides. One of her questions was whether they believed themselves to have suffered any ill-effects from agrochemicals. Despite their conviction that pesticides are an essential part of modern farming, 41 farmers admitted to suffering ill-effects, that is one half of the farmers in her sample believed that they had become ill as a result of using pesticides. (And remember that by no means all symptoms of chemical poisoning will be recognised as having been caused by pesticides, so the real figure for pesticide related illness is likely to be even higher.)

So why do the official figures remain so reassuring? One reason was neatly identified by another of Dr. Tait's questions. When asked whether they had reported the suspected side effects of pesticides to the Health and Safety Executive, or anyone else, only one of the farmers had bothered to do so. And this man sent in a complaint because his wife had been affected by another farmer's spraying! It appears that there is virtually no reporting of pesticide ill-effects by users. The official statistics are not worth the paper they are printed on.

A recent survey by several British trade unions shows up these discrepancies. In *Pesticides: the Hidden Peril*, a questionnaire of almost 300 users is reported to show that:

- 50 per cent of users reported that they had suffered the 'classic' symptoms of headaches, sickness and sore throats as a result of chronic, low level exposure through using pesticides.

- This occurred despite the fact that over 80 per cent were supplied with and wore protective clothing.

- However, 20 per cent said that their protective equipment was ineffective and not properly maintained.

- 38 per cent of the respondents had received no training on the health and safety aspects of pesticide use and the vast majority of those who had received training had less than a day's worth of instruction.

- 5 out of the 13 most frequently mentioned chemicals have been banned or severely restricted in one or more countries.

- 9 of the commonest chemicals are suspected of having chronic effects (either causing cancer or birth deformities) and 11 are irritating. All are one or the other, many both.

The 13 most commonly mentioned chemicals were: amitrole, atrazine, 2,4-D, dichlorbenil, dimethoate, glyphosphate, lindane, malathion, mecoprop, nicotine, paraquat, propyzamide, simazine.

• How pesticide users become contaminated

This being the case, just how are pesticide users most frequently contaminated? A major cause of contamination is undoubtedly the lack of knowledge about health risks among many people handling agrochemicals. Advertisements present pesticides as safe; sales representatives understandably minimise the hazards associated with their products; and many agricultural 'experts' have been educated in an atmosphere of blind faith in the power and safety of the chemical spray. It is not surprising that many users simply don't register the dangers involved.

Sometimes the degree of plain ignorance is frightening. A senior staff member at one of the major agricultural machinery manufacturers in eastern England told me that a common way of spray users becoming contaminated is through clearing a blocked spray nozzle by putting it into their mouth and blowing through it. They might almost as well drink some pesticide.

Another example came up in a public meeting about pesticides in Sussex. During a fairly heated debate about hazards, one crop-spraying pilot cheerfully admitted to me that he had been drenched by all the chemicals he applied at one time or another, without apparently suffering any health damage. He was, to his own way of thinking, a responsible man and stressed that he would do nothing which he thought would be harmful to his two-year-old daughter. Maybe so, but many of the chemicals he probably would have been using are known to have chronic effect on people's health in the long term. When I have mentioned this example to people in the agro-chemical industry they have said, well of course he should never have done that, it was breaking all the safety regulations, we cannot be held responsible for individual stupidity, and so on. Well of course. But the overall impression of chemical safety put across by the industry was sufficiently powerful that a responsible young pilot had missed out on some of the most basic points about spraying hazards. And he is by no means unusual.

The only way to tackle such ignorance is by proper, professional training for all pesticide users. This is already a legal requirement in some countries, including the USA. There, trainees learn not only about looking after their own health, but have to study details of a wide range of environmental effects and health implications for the general public, etc. as well. Although Britain now has plans to introduce mandatory training, the details of exactly what will be taught are still unclear. And, as present proposals have it, training will only be a legal obligation for people under 25, the assumption being that older users already know enough through experience. The examples given above show such reasoning to be a nonsense. Current plans mean, at best, that there are years ahead where many users with almost no grasp about the dangers of agrochemicals are still allowed to use them on a large scale in public places. At worst, the training schemes will never be properly organised, and users will continue to rely on advertisements and enthusiastic professional advocates for their safety 'advice'.

In any case, we would be wrong to assume that training could solve all the problems of pesticide danger to operators. It is often claimed that there is no danger if pesticide is applied properly. Well publicised experiments carried out by the chemical industry have shown what everybody knew already; that spraying carefully on a flat field, wearing protective clothing and in conditions of little wind, results in little visible contamination of sprayers.

In the real world of a working farm, or in upland forestry plantations, usage is seldom so simple. Sprayers will be working under pressure of time and may have to apply pesticide in windy conditions if the weather so decrees. Forestry workers will often be spraying on steep slopes and clambering through brushwood. People will get tired, protective clothing will get torn. It is expensive to replace, so tends to be used for longer than it gives adequate protection. If ordinary clothes are worn, the pesticides can contaminate the material and increase total exposure. Anyone who has used a backpack sprayer for more than a couple of hours will know how exhausting it can be, especially in hot weather. Sweating in a noisy tractor cab and negotiating the sprayer over difficult terrain for hours on end is little better. People lose concentration and accidents happen. The fact that operators suffer far more from short-term health effects than official figures suggest is no longer open to doubt. Long-term health effects are not even mentioned.

New evidence from the USA suggests that, even if the users follow instructions to the letter, they are still at risk. Experiments have now shown that fine spray droplets can penetrate straight through some of the rubberised or plastic-coated 'full protective suits' worn to apply the most toxic agrochemicals. Someone dressed up like a moon explorer can be almost as much at risk as an unprotected person. In fact, the risk will probably be greater, because full protective gear is normally only worn for the most hazardous chemicals, and, believing themselves to be completely safe, users will probably be less particular about avoiding pesticide spray. However you look at it, the pesticide user is amongst the highest risk categories for people suffering toxic effects from agrochemical pollution.

• Spray drift

In the rich, loam soils of Lincolnshire, near the small town of Spalding, a couple run a smallholding famous for the quality of its produce. (Names have been omitted from this example because it is still subject to a court hearing.) Indeed, the vegetables they grew were judged to be so good that lettuces were supplied to the Royal kitchens. In 1980, a neighbouring farmer sprayed his crop with fungicide, in conditions that were too windy for safe application. A cloud of pesticides drifted from his land and through the neighbouring smallholding, covering crops, garden and house in a fine chemical mist. Within days, thousands of pounds worth of vegetables became distorted and were destroyed for commercial purposes. The profits for the year had disappeared at a stroke.

The growers immediately contacted their local Agricultural Development and Advisory Service (ADAS) official, who confirmed that the damage was due to spray drift. Shortly afterwards, they had a visit from a local National Farmers Union representative. Far from being supportive, he had come in the hope of persuading them not to make a complaint. Complain they did, but with little success because the farmer denied responsibility. Faced with financial ruin, they went ahead with a court case to sue for damages, but in the years that have passed since, nothing has been resolved. The farmer still denies responsibility. The growers are still out of pocket to the tune of thousands of pounds, and have had to lay off their worker. When I visited the farm in 1984, they could still show me the passage that the spray drift took by the stunted trees and bushes in the garden. They both claimed to have suffered illness as a result of breathing in the fungicide. Without a doubt, the period of strain has taken a toll on their mental and physical health.

The growers described above have had a large portion of their lives ruined by one thoughtless (or selfish) application of pesticide. Despite coverage in the national farming press and a sympathetic and helpful MP, they have still failed to get any action after years of effort. As in many other cases of spray drift damage, they have had to deal with official indifference, a hostile union, professional suspicion and incredibly slow court procedures. Like other sufferers, they have found that being in the right does not necessarily help that much.

'Spray drift' means the uncontrolled drift of agrochemicals away from the intended target area and on to roads, fields, gardens and parks. Drift of hazardous chemicals threatens crops, livestock, wildlife and, of course, people. The term 'spray drift' is used fairly loosely to describe three different effects:

spray drift or **drop drift**, caused by small droplets of agrochemical spray being carried away from the target area by air currents.

vapour drift or **volatilisation** (evaporation) of chemical already on plants or in the soil, followed by drifting in the wind.

blow of chemicals spread along with soil particles by stronger winds.

Whilst no one denies that spray drift problems occur, the official attitude is that these are very small compared with the total use of pesticide; an acceptable risk where any damages can usually be settled out of court. Figures for reported cases of damage in 1983, for example, suggest that there were only 153 incidents involving aerial spraying throughout Britain. No particularly significant problem, you might think. Yet when I spoke at a public meeting about pesticides in Louth, in Lincolnshire, during August 1984, well over 40 people in the audience claimed to have suffered problems from pesticide drift as a result of aerial spraying. Many had detailed stories of being contaminated time and again. Some people were persistently sprayed in their own gardens, others on roads and public footpaths. Unless over a quarter of the British cases just happen to occur in Louth, the official figures are a gross misrepresentation of the real situation.

The inadequacy of the British government's statistics about pesticides was neatly illustrated in a debate in the House of Commons during the passage of the 1985 Food and Environment Protection Act. The government spokesman claimed that problems of spray drift contamination were extremely rare. Within minutes, two of his Conservative colleagues had jumped to their feet with stories about their own experiences of pesticide drift. One had been sprayed while driving his car and another, Trevor Skeet, had been caught in pesticide drift from a crop-spraying plane while standing in his own garden in Bedfordshire. As in the case of user contamination, official figures about spray drift problems bear little relationship to reality.

Damage to crop plants

The best known effect of spray drift does not directly affect humans, but is damage to other crop plants, commonly caused by drifting herbicides. Some herbicides are general defoliants, used to clear away weeds before planting or on pavements and verges, and these will damage any plant they come into contact with. Others are more specific, but will almost always affect a range of species. A common problem occurs when herbicides used to clear weeds from amongst grain crops drift over leafy vegetables. This is especially likely where arable farms and market gardens are mixed together in one area, such as the Vale of Evesham in Worcestershire. In 1981, a quarter of a million pounds worth of claims were outstanding in the Vale because of spray drift incidents. A research paper by the World Resources Institute, based in Washington DC, reported estimates of annual damage from spray drift worth 11 million dollars in Texas alone.

The reality of spray drift damage is not open to doubt and the Agricultural Development and Advisory Service (ADAS) has even produced a guide to herbicide damage which includes details about recognising spray drift. The British Crop Protection Council produced a major report on spray drift and efforts have been made to inform farmers about the problem, including distribution of leaflets to thousands of farms. But the problems continue.

Livestock

It is far more difficult to get official confirmation that livestock (or people) have been affected by drift. This would mean admitting that the pesticides were dangerous. Nonetheless, well documented cases of livestock damage from spray drift already exist.

Geoffrey Helier, a smallholder in the Quantock Hills of Somerset, saw spray drifting out of a Forestry Commission plantation over an area of grassland where his 180 Border Leicester sheep were grazing. When questioned, forestry workers told him that the pesticide being used was 2,4,5-T and several of the men were refusing to use it themselves because they feared it was carcinogenic. Later, Helier's ram, 25 ewes and 59 lambs died. Many more ewes aborted and the surviving lambs were sickly. The Forestry Commission denied that drift had occurred, despite Helier's eyewitness testimony. His vet failed to test for poisoning in the autopsy. By the time it was possible to retest the sheeps' corpses it was too late to carry out the analysis. Nothing could be proved and Helier received no compensation. Still worse, during his efforts to complain the story of the poisoning had spread around the district and he was unable to sell perfectly healthy sheep the following year, because local farmers feared that they might be infected.

Recently, several successful claims for damage to cattle and sheep have been made against farmers who have caused spray drift. However, Helier's case shows how hard this can be in practice. Detecting one of a range of chemicals is bad enough, but proving where it came from (especially when several sprayers are operating in the district) is often impossible.

Bees

One small 'livestock' known to suffer badly from spray drift, and from pesticide use in general, is the honey bee. Bees are important both as honey makers and as pollinators of crops and fruit trees; many hives are rented out to farmers for the latter purpose. Bees can be caught directly in spray drift or contaminated when they feed on treated plants. Beekeepers have come to recognise a pesticide 'hit' by the hundreds of dead bees found lying outside a hive. Even those still fit enough to enter are repelled by the guard bees, because the chemical gives them an unrecognisable smell.

The recently developed practice of spraying oilseed rape early in the season has made the bees' plight even worse. Many hives are lost every year. For example, in 1983 two beekeepers near Andover in Hampshire lost all the flying bees from 32 hives when a farmer sprayed oilseed rape with Lindacol, containing **gamma HCH**. Although beekeepers often organise their own early warning system in cooperation with local farmers, and keep hives shut down during spraying, losses still run very high. (And, of course, damage to bees also means damage to other insects. Wildlife effects are examined in detail in a following chapter.)

• *Causes of spray drift*

Given that spray drift is undoubtedly an important problem, how exactly is it caused? As is so often the case when looking at pesticides, we find a mixture of human and technical failings which combine to cause a major problem.

At present about a thousand million gallons of pesticide-containing spray are released on to the British countryside every year. We are told that these quantities are essential to our agriculture and our well-being. Yet in a lecture to the Royal Society in 1977, I. J. Graham-Bryce stated that less than one per cent of insecticide applied is actually effective in killing pests. Virtually all the insecticide is wasted. Even with herbicides, where efficiency is far greater, 70 per cent or more is routinely 'lost'. Within these colossal inefficiencies lie many of the answers to how spray drift occurs.

It is an apparent paradox that, whilst chemicals used in farming are now increasingly complex and sophisticated, the equipment with which they are applied to the crop has changed little in fundamental design throughout the whole of the twentieth century. Conventional hydraulic nozzles simply force the liquid out under high pressure. The resulting spray is made of droplets of widely different sizes, with the largest containing something like a million times more chemical than the smallest. Both large and small droplets are of little use for pest control. Large droplets tend to bounce straight off the plant or, if they do lodge in vegetation, contain far more chemical than is required for pest control. The smallest droplets, less than 100 microns in diameter, usually drift away whatever the weather conditions. Only a small proportion of the chemical is released in a form suitable for treating the plant.

Some recent innovations have, if anything, made the problem even worse. Wider spray booms are now used to increase the so-called efficiency with which spray is applied, by increasing the area treated at any one time. But these wide booms bounce more, throwing droplets higher into the air and thus increasing drift. High spraying pressure also has the effect of adding to the amount of drift.

Yet more efficient methods of applying spray already exist. Controlled droplet application (CDA) methods attempt to regulate droplet size to a narrower range, which has more chance of reaching the target pests and controlling them. One common method of CDA is to use a spinning disc or mesh cage to produce a relatively constant sized droplet. The sprayer can be calibrated to produce a droplet size suitable for a particular set of biological and climatic conditions, so as to maximise the effective amount of spray. The system is also known as the 'dial a droplet' method. (CDA and hydraulic nozzles are illustrated opposite.) Farmers using CDA claim to have cut their pesticide requirements to a fifth of what they used previously, and even greater savings will become possible as the technology is further modified.

Controlled droplet application is already extensively used in parts of Europe. However, introduction of CDA into Britain has been considerably slowed by entrenched opposition from some quarters. Articles in the farming press have pointed out

A typical spray pattern from a conventional nozzle

A typical spray from a CDA nozzle

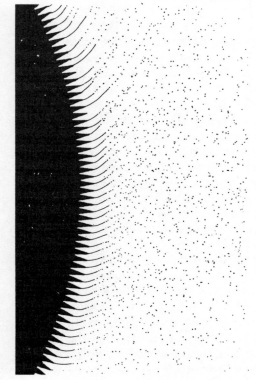

its 'inefficiency', sometimes using rather outdated test results. Little money has been made available for research. The Soil Association's book, *Pall of Poison* concluded 'reduced quantities of chemicals used for spraying means reduced profits for chemical companies. The farmer, far from being the villain, is just another victim of corporate efforts to maintain and expand a lucrative market'.

Now that the issue has been blown open and discussed, changes may come about in the near future. And the chemical industry is apparently developing new chemicals in preparation, suitable for use in CDA, which will be sold at higher prices to ensure that greater efficiency for farmers does not reduce the profits from agrochemicals.

● *Vapour drift*

Vapour drift is another aspect of the spray drift problem, and has produced a different response from the various interest groups involved. Vapour drift occurs when particularly volatile herbicides (especially ester formulation hormone sprays) evaporate some time after being applied to crops. This evaporation can take place several days after the herbicide is applied, and this is even more difficult than droplet drift to trace back to any particular source.

Although vapour drift has caused extensive damage to market gardens in some places, the biggest sufferers have probably been the farmers themselves. Vapour drift frequently damages the crops growing directly beside the treated area. In Britain, this was especially noticeable in the case of oilseed rape growing next to grain.

Travelling through England in the early 1980s, it became common to see fields of wheat or oats alongside a bright yellow field of oilseed rape, with a swathe of flowerless rape showing up at the boundary between the two. Herbicide used in the grain fields was evaporating and drifting over the fence to kill off the rape. Faced with very obvious losses, the National Farmers Union moved quickly and imposed a voluntary ban on volatile herbicides in March 1985.

• *Aerial spraying*

In July 1984, seven part-time firemen from Welwyn, in Hertfordshire, were called out to tackle a haystack blaze near Kimpton. As they fought the fire, two crop-spraying planes were working in adjacent fields, one spraying insecticide and the other fungicide. The men were enveloped in a cloud made up of the two sprays and, almost immediately, had to be rushed to hospital with breathing difficulties and streaming eyes. Although they were let home after 24 hours of X-rays, eye irrigation and nose washes, they still felt ill a week later and the fire station had to be temporarily closed in consequence.

Aerial spraying now makes up only two to three per cent of British pesticide application, but it generates a far larger proportion of complaints from the public. In other countries, and especially in the Third World, it is still a very important method of pesticide application and treatment of large areas. Inadequate rules, a cavalier attitude amongst many pilots and the inherent dangers of spraying from a fast-moving plane all add to the hazards posed by aerial crop spraying. Despite recent tightening of legal restrictions on crop spraying in Britain, there was strong all-party support for a complete ban on aerial spraying of arable crops. However, the government rejected this option.

Most pesticide regulations in Britain fall naturally under the auspices of the Ministry of Agriculture, Fisheries and Food (MAFF). However, important sections of legislation concerning aerial pesticide application are controlled by the Civil Aviation Authority (CAA), which is part of the Department of Transport. The issue apparently causes much inter-departmental wrangling within the confines of the Civil Service, but strenuous efforts by MAFF to gain control of aerial spraying have so far failed.

Britain has some of the weakest controls over aerial spraying in Europe in terms of minimum permissible spraying distances from roads and dwellings. But even these look far better on paper than they work out in practice. Officially, all pilots have to be certified by the CAA under the Aerial

Applications Certificate; a fairly detailed set of guidelines about what can and cannot be done by crop spraying planes. This includes requirements for site reconnaissance; informing neighbours about spraying before it takes place; avoiding flying over livestock; setting minimum distances from major roads and dwellings; and limiting the number of chemicals used to just over 70 individual formulations and mixtures, some of which are only cleared for use as granules.

In 1984, only 100 aircraft were registered for aerial spraying in Britain. However, it is known that additional planes were also used in clandestine and illegal spraying operations in some areas.

> In June 1984, a crop-spraying plane was operating on a farm near Tewkesbury in Gloucestershire. It contaminated a neighbouring field, belonging to Stella Hill, and sprayed pesticide onto sheep, some valuable brood mares, and Mr and Mrs Hill themselves. Soon afterwards, one of the sheep became so ill that it could no longer walk and the horses were also extremely unwell, suffering acute diarrhoea. The Hills' vet needed information on the nature of the chemical used, but the farmer refused to give this, even when contacted officially by the local police. As Stella Hill had noted the plane's registration number, the police took up the case directly with the CAA. When the plane was traced to an address at a local farm, it turned out that the firm concerned had disappeared overnight a year previously and were certainly not registered for pesticide application. The case has never been satisfactorily resolved. The firm was alleged to have regularly broken spraying regulations in the past.

However, it is not only illegal operations which cause problems. The short operating season, high risks and the uncertain nature of the trade have attracted a high proportion of 'cowboy' operators who, although legally registered, apparently have little regard for the finer points of flying safety. Once again, the number of accidents and incidents of illegal use appear to be far greater than reassuring pronouncements from the industry and government suggest.

Whilst carrying out research for a Soil Association report on aerial spraying in 1984, I made a fairly cursory survey of incidents which actually reached the columns of local papers, which is probably quite a small proportion of the total. The 45 British cases which I identified included examples from 17 counties of England and Wales. Gardens, sports fields, villages, roads and individual houses were all sprayed. 20-30,000 bees were killed in one incident in Boston, Lincolnshire; fish were killed in a pond near Durham; cage birds died after pesticide drift reached a house in Saltburn, Yorkshire; potatoes withered in gardens in another Yorkshire village; motorcyclists were sprayed in Staffordshire and Lincolnshire, the latter suffering facial burns and dramatic changes to blood plasma levels; and a policeman was contaminated whilst in the process of answering a complaint about aerial spraying, and suffered a rash and sore eyes, nose and throat as a result.

When a helicopter sprayed the village of Amesbury in Hampshire, Salisbury District Council described it as 'sheer bad flying and gross carelessness'. After a near fatal crash between a crop sprayer and a military aircraft, a Lincoln-based RAF officer commented that spray pilots 'think they have a God-given right to spray where they like'. The list could go on. In the same year, Friends of the Earth compiled a *Pesticides Incidents Report* containing over 200 case studies, many of which included aerial spraying.

It is not hard to see why. Pilots are locked up inside small cockpits, flying long distances (so that they are less able to judge local weather conditions), working against the clock and releasing pesticides at high speeds. All classic components of a high industrial risk activity.

• *Why are official figures of accidents so low?*

The examples quoted above illustrate a general point, which is true of almost all pesticide use and abuse. I have repeatedly pointed out that only a fraction of cases ever get reported at all. One important reason for this is that it is often extremely difficult to report a pesticide incident. And if you do, it is all too easy for people to cast doubt on your story. Over several years of talking to people who have suffered pesticide problems, I have been told time and again of victims being rebuffed by officialdom and red tape when they try to complain about pesticides.

Stella Hill, in the case described earlier, was told by several different police officers that it was none of their business, until she found a sympathetic sergeant to listen to her story. And if the complainants do persist, they come up against the professional lobbyists of the industry, who are adept at brushing off criticism and defusing rightful anger. When you add on the fact that many people living in the country will be very unwilling to report a neighbour, who could be a personal friend, or perhaps employ one of their relatives, the ways of filtering out complaints become even more complete. In practice those who do complain are often outsiders; professional people who commute into the city by day and who know the law and are not in awe of any local landowners.

Take the collapse of the Welwyn firemen for example. A couple of years previously a crowd of Nottingham schoolchildren became ill in the same way after attending an outdoor sportsday. In 1984, a girls' jazz band collapsed while marching through Rawcliffe Bridge in Yorkshire. Crop spraying was suggested in both cases, and in the latter incident stains of some liquid were found scattered over the sufferers' clothes. The HSE found no traces of pesticide. Both cases remain unsolved, and it is now too late to prove anything one way or another. In both the police suggested mass hysteria as the most likely cause. Well perhaps this is true, but it does illustrate an interesting point about attitudes to pesticides. Say that a crowd of schoolchildren or young girls have, like the heroine of a Victorian novel, submitted to a case of the vapours and you stand a good chance of getting away with it. Say the same about a bunch of firemen and you are more likely to get a punch on the nose.

On 13th September 1984 the *Oxford Mail* reported a prosecution brought by the Health and Safety Executive against Jeffrey Jones of Manor Farm, Upper Hayford, who was fined £100 plus £50 costs after 'admitting failing to ensure that other people were not exposed to risks to their health and safety'. People working in workshops bordering Mr Jones's land 'noticed a smell, like burning rubber, and then suffered choking sensations' while he was spraying by tractor. Mr Jones's defending counsel said that it would have been difficult to warn neighbours when spraying the fungicide/insecticide mixture because there were 'hundreds of them'.

To protest about aerial spraying it helps to have the right background so you will know how to complain effectively and have a proper hearing. It also helps to be the right age and sex, and to have witnesses. And to be persistent. Some advice about how to complain about pesticides, whether or not you fit in to these categories, is given at the end of this chapter.

● The chemicals used in aerial spraying

We have established that aerial spraying is inherently dangerous and that it attracts a fair quota of illegal or semi-legal operators. But there are also inevitable dangers of aerial spraying in Britain because of the types of pesticides which are cleared for use here. Although these are supposed to be rigorously assessed and the more hazardous chemicals banned, the reality is that many of those passed have big question marks hanging over their safety. The Appendix List illustrates some of these dangers. They underlie the decision by both the Labour and Liberal Parties to call for a ban on all aerial spraying on arable land, where the technology now exists to replace it completely with safer, land-based spraying systems.

● The problem of very small droplets

In the section on droplet drift we looked at the losses and inefficiencies which result from the production of both very big and very small droplets. The small drops are liable to drift whatever the weather conditions.

Until recently, it was assumed that the proportion of very small droplets (i.e. less than 100 microns) was comparatively low. Recent experimental evidence suggests that this is not the case. Scientists at the Overseas Pest Research Centre at Ascot have utilised a new laser measuring system to work out the proportion of pesticide released in the various different droplet sizes. Far from finding an insignificant number of small drops, they measured up to 20 per cent of the pesticide being released in particles with a drop diameter of less than 100 microns. These droplets are extremely liable to drift, and can travel very long distances in the atmosphere. If water is used as a dilutant it can evaporate away within seconds of being released from the sprayer, leaving a particle of 'pure' chemical which can drift almost indefinitely in the atmosphere.

There are no signs that the situation will necessarily improve in the future. Even if CDA systems become more widely adopted, there is now an increasing use of ultra low volume (ULV) spraying techniques, where the sprayer chooses very small droplets. Although proper use of ULV can dramatically reduce pesticide requirements, as is the case with all CDA systems, slight miscalculations could result in very substantial releases of tiny droplets.

Again, another recent innovation is 'electrostatic spraying', whereby droplets are electrically charged so that they are attracted to the vegetation to maximise efficiency. However, ULV sprays are used for electrostatic spraying, and there has apparently been no research to look at the losses of small droplets with this system.

Let us be optimistic and halve the proportion of very small droplets produced from the Ascot figures, to 10 per cent of sprays released. Remembering that roughly a billion gallons of pesticide containing spray is used in Britain every year, this would mean some 100 million gallons of pesticide solution being released as extremely driftable droplets over rural areas of the UK.

We simply do not know what the implications of this are for human health or the environment. The very small chemical particles may stay in the atmosphere almost indefinitely. Dust particles of similar sizes have been found in the stratosphere. Small amounts of a whole range of pesticides are being breathed right now by people throughout agricultural areas.

Many of these chemicals are extremely dangerous and require that users wear special protective gear. We have virtually no idea what long-term exposure to very small quantities, and to mixtures of many different types, of agrochemicals mean to human health.

Continuous low level contamination is virtually ignored in discussions about pesticide hazards. Yet it was people within the chemical industry who first drew the attention of the Soil Association to the possible health implications of this aspect of spraying. The hazards of spray drift may be considerably more wide ranging than we have recognised until now.

• What to do about spray drift and other pesticide damage

'The somewhat complicated control machinery . . . inevitably presents some difficulties to members of the public who wish to complain, or to obtain advice, about a spraying incident. We have seen accounts of the dealings of members of the public, or even local authorities, with the official bodies involved which suggests a lack of clarity in responsibilities for dealing with incidents . . . We consider that there is a need to review the arrangements for dealing with incidents, with the aim of clarifying responsibilities and lines of communication so that advice is readily available to people who believe that they have been subjected to a chemical spray.'

The Royal Commission on Environmental Pollution Seventh Report: *Agriculture and Pollution*, 1979.

Many people who suffer pesticide damage also find themselves facing additional problems when they complain about what has happened. Police and health authorities are often unaware of their statutory obligations. So questions about health effects are shrugged off with bland assurances, and victims who have lost money through pesticide damage often wait for years to get any compensation.

This section summarises how you can recognise pesticide

damage, who to contact for more information about agricultural and health problems, where to register a complaint, and how best to protect yourself against pesticides. Hopefully, it will also help people let the authorities know when they are unhappy about the way pesticides are used.

Crops

Growers can claim compensation against spray operators for drift damage to crops. Unfortunately, proving the existence and cause of damage is not easy. Symptoms usually appear several days after drift occurs and often look similar to the effects of mineral deficiency or disease; by this time it may be impossible to identify pesticide residues in the soil. Taking someone to court is a long and risky process and the odds are often stacked against the victim. However, a regular procedure for claiming has developed, which maximises the chances of a successful claim.

The grower's strongest position is when the spray drift is discovered at the time of its occurrence. Drifting spray is sometimes seen, but is more likely to be detected by smell or taste. If this happens you should immediately:

1. Follow the drift upwind to find the source.

2. Inform the sprayer that they (or their employer) will be held responsible for any damages occurring as a result.

3. In the case of contract spraying, also tell the owner and suggest that they contact their insurer at once.

4. Immediately write down details and record a census of all crops which might have suffered damage. Photograph crops at the time.

5. Confirm the occurrence in writing to the land owner within 24 hours, heading the letter 'Formal notice of claim'. Deliver by hand if possible.

6. Contact ADAS (or the Scottish equivalent) for expert advice about possible damage. ADAS will examine crops to determine both if they were healthy before drift occurred and if they suffer damage afterwards, and will state the facts in writing, but will not advise about litigation or assess losses. They can also test soil samples for pesticide residues if contacted early enough.

7. An independent consultant, land agent, NFU secretary, valuer or solicitor should be asked for advice on crop loss assessment and compensation proceedings. (Some assessors have a reputation for favouring farmers against growers so try to check anyone out first with previous claimants.)

8. In the case of insurers wanting to see the crop, they must be allowed access as soon as possible, ideally in the presence of an independent witness (but not from ADAS).

Going through a long procedure is difficult for a working grower and may seem like wasted time. If there is just a small amount of damage to a private garden, you may simply want to complain vociferously and not bother to take the matter further if this gets no response. But failure to follow these steps will make claiming any damages much more difficult if your crops are affected.

Detecting crop damage

To get any compensation you will need to identify crop damage. Unfortunately, crops differ in their susceptibility to and in their symptoms of drift damage. Crops can suffer sub-lethal doses which affect yields but are not detectable to the un-trained eye. If damage is severe, soil and tissue analysis can sometimes identify the chemical. However, a detailed know-ledge of both the chemicals and crops is usually needed to make any judgement. The following summary is just a primer on the type of effects to expect.

Crops sensitive to growth regulators

Glasshouse crops   tomatoes, cucumbers, peppers, lettuce, courgettes, bedding plants, seed beds of all kinds, celery, roses, chrysanthemums and pot plants.

Outdoor crops   lettuce, marrows, beetroot, peas, beans (broad, french and runner), brassicas (all kinds), mustard, turnips, swedes, parsley, nursery stock, decorative trees and shrubs, herbaceous perennials.

Trees broadleaved trees in hedges are usually good indicators of drift passage, showing leaf malformations, petiole twisting (this is the slender shoot joining leaf to twig), stunting and shoot kinking. Most conifers are not as sensitive.

Wild herbs can also show where drift has been near hedges or semi-natural areas.

Certain common pesticides also produce distinct and recognisable symptoms. Some of these are illustrated in the table below:

Chemical	Symptoms
Paraquat/diquat	Circular, chlorotic spotting on leaves, followed by paling/yellowing. Some broad-leaved species quickly wilt, suffer interveinal discolouration and leaf edge blackening, browning and top-growth death.
Amintriazole	Symptoms take up to two weeks. Plants wilt, leaves pale then turn white (sometimes pinkish), starting at growing points and working down leaf.
Glyphosphate	Symptoms slowly develop over three to six weeks. Stops growth, growing points yellow, wilting occurs, die-back goes gradually down the plant.
Contact herbicides	Scorching, yellowing and death of soft tissues, especially at growing points, flowers, young leaves and fruits.
Chlormequat	Stunting and sometimes almost complete suppression of growth.
'Hormone growth regulators'	Complete disruption of the growth processes in many plants. Can also reduce the storage life of crops.

For a more detailed description of herbicide damage, see the ADAS publication *Diagnosis of Herbicide Damage to Crops* ADAS/MAFF Reference Book 221; 1981

• *Making a claim*

Most spray drift cases are settled out of court. If the procedure outlined above is followed, and an early agreement about the extent of damages is reached, spray operators are often willing to pay up quickly to avoid ill feeling. However, it is often worth getting a lawyer to write a letter to the spray operator or farmer to speed up the process. Claimants should remember to:

- wait several weeks before assessing damages; early claims are almost always too small;
- charge the revenue lost rather than just the profit, i.e. include labour time, capital, running costs, etc.;
- include reduction in yield, shelf life, growing time, quality and carry-over charge in the case of perennial crops.

Unfortunately, there are a small, but increasing, number of people prepared to fight through the courts to avoid payment, or who try to get away with offering ridiculously small sums in compensation. In these cases, the plaintiff is left with the burden of proof of damage and drift source. The last factor, especially in areas where many farmers are spraying at one time, can prove a serious stumbling block for claimants.

- Check that your solicitor has a reputation for handling cases quickly. Picking the wrong person can lead to (literally) years of delays.

Fortunately, establishing drift is 'on the balance of probabilities' instead of absolute proof positive. This makes it far easier to establish a verifiable case in court. Common ways of finding out where drift comes from include:

- following drift back by tracing damage to trees and wild plants;
- establishing where the prevailing wind was blowing from at the time (via the Meteorological Office, London Road, Bracknell, Berks.);
- claiming against all offenders in the case of there being many possible sources.

In the courts, the most commonly cited legal precedent is Rylands versus Fletcher. This is a classic nineteenth-century ruling that 'operations carried out on land which inadvertently affect a neighbour are liable to compensation'. Going to court is very much second best as an option. It takes a long time to get a writ served (at least two to three years waiting list) so the claimant has to carry costs until then. Some insurers are prepared to try to scare claimants away; for example in one recent case the defendants held back a vital witness, who had also been spraying on the same day (thus destroying the claimant's proof that the defendant was the culprit in spraying him) until the last minute, so that the claimant incurred heavy costs as well as losing his claim.

Livestock

Livestock will not usually be immediately or obviously poisoned unless pesticide application rules have been broken. This means that sprayers are even less likely to admit responsibility. Virtually all claimants of livestock damage have had to fight hard for their compensation. Livestock can be affected in a number of ways:

- by being caught directly in spray;
- by eating plants which have been sprayed, either being poisoned by the pesticide, or eating poisonous plants which the animal no longer recognises because the chemical has given them a different taste and smell, also by eating granules of insecticide (like slug killer);
- by contact with freshly sprayed vegetation, such as crops or pasture treated with paraquat, and being poisoned through skin contact or vapour. (This can also affect pets.)

The local vet should be able to advise on the possibility of pesticide poisoning, although this is not always the case.

Bees

Beekeepers suffering a kill should send at least 200 dead bees, plus relevant information (including type of chemical if possible) to the ADAS National Beekeeping Unit, Luddington, Warks., which monitors poisoning. If pesticide damage is proved, beekeepers can claim compensation. In 1984 a judge set a precedent that bees were more like children than burglars and that 'their foraging behaviour is seldom if ever harmful'!

He said farmers have a duty to warn local beekeepers before spraying. But beekeepers then have a duty to confine or relocate their bees. This puts beekeepers in a far stronger legal position and they should tell all neighbouring farmers that they wish to be informed about spraying.

Recognising bee poisoning Bees affected by pesticides will either die in the field or make their way back to the hive, where they will often be refused entrance by the guard bees because the chemicals give them an unfamiliar smell. Large numbers of dead bees on the ground and, especially, around the hive, are indicative that some form of poisoning may have taken place.

Beekeepers can also take a number of steps in cooperation with farmers to minimise damage. They should urge farmers to:

- spray in the morning or evening when fewer bees are flying;
- avoid spraying flowers when more than ten per cent are in blossom;
- use less toxic herbicides. Notes on toxicity to bees are given in *Approved Products for Farmers and Growers* (see Books, p. 170);
- give beekeepers at least 24 hours warning of spraying within two miles of their hives.

The efficiency of the local, voluntary bee liaison officer is of vital importance in building up a good rapport with farmers and ensuring a constant flow of information.

Wildlife

Damage to crops and bees shows that wild plants and insects may also have been poisoned. Birds and mammals suffer from toxic effects themselves and populations also decline when natural foods become scarce. The Nature Conservancy Council (NCC) conservatively estimate that ten per cent of Sites of Special Scientific Interest (SSSIs) are harmed by pesticide sprays every year.

At present there is very little legal protection for wild habitats, beyond an obligation to inform the NCC before aerial spraying near to a National Nature Reserve or SSSI. However, spray drift does undoubtedly damage wildlife, so owners of land dedicated to the conservation of wildlife should try to get pesticide use prohibited from their borders.

Birds found dead where spray has been used should be reported to the Royal Society for the Protection of Birds, The Lodge, Sandy, Beds. (0767 80551). Dead mammals and large kills of reptiles, amphibians and insects are of interest to the Royal Society for Nature Conservation, The Green, Nettleham, Lincs. (0522 752 326). Local representatives of the Royal Society for the Prevention of Cruelty to Animals (national office), The Causeway, Horsham, West Sussex (0403 64181) may also help, particularly if you suspect a deliberate poisoning. Carcasses need to be labelled with the date and place of collection and deep frozen in case they are required for later analysis.

Bats are at particular risk when timber is treated with organochlorines like Lindane and Dieldrin and also Pentachlorophenol against woodworm. These can remain toxic to bats for literally decades and have disastrously reduced numbers in some areas by killing colonies nesting in roofs. Less toxic alternatives now exist and so it is worth asking about the safety to bats and other creatures when making a purchase. All bats are now subject to special protection under the Wildlife and Countryside Act.

Human health

Symptoms of immediate pesticide poisoning are described in Chapter one. A method of getting an indication of pesticide poisoning is to have a blood test within 24 hours (or at most 48 hours) after suspected exposure. High levels of pesticides in the blood strongly suggest that they are the cause of the symptoms. Unfortunately blood testing is not usually done by the National Health Service at present, unless you are admitted into casualty with severe effects. One private hospital willing to carry out blood tests is the Nightingale Hospital, 19 Lisson Grove, Marylebone, London NW1.

● *Fighting back*

Trying to fight back against illegal or dangerous use of pesticides can be an exhausting waste of time unless it is approached in the right way. The section above should have provided an introduction to how to recognise and prove pesticide damage, but knowing how to use this knowledge is also important.

If spray damage occurs, contact police, health officials and your doctor if necessary. Their response will be varied. Some are sympathetic and helpful, while others don't understand the law or are reluctant to get involved. The police *are* responsible for complaints about spray drift, but we have come across several cases where local police refused to get involved and claimed it was not part of their duties.

Whilst individuals can obtain damages, it is more difficult to question the overall safety of local pesticide use. Someone working alone will probably just be labelled a crank. Most effective protests have involved a number of people cooperating by watching for illegal spraying incidents and pooling skills. Once a few people make a fuss they attract other help, sometimes through anonymous phone calls, and working together also keeps up enthusiasm.

If enough of you feel there is a serious problem, don't be afraid to take the matter further. The media are often helpful, and it is worth getting to know newspaper reporters and people at radio stations who are looking for news. Two Pluto Press handbooks, *Using the Media*, by Dermott McShane and *Get It On. . . Radio and Television* by Jane Drinkwater are useful guides on how to maximise your impact.

One way of determining the strength of local feeling is by holding a public meeting. A couple of speakers and contributions from the floor make a good combination, especially if several spray drift victims are ready to get the discussion started. A meeting in Lincolnshire produced 40 people who had experienced spray drift. This is almost a quarter of occurrences nationally according to official sources! Another Pluto handbook *Organising Things: a Guide to Successful Political Action* by Sue Ward gives information about setting up public meetings.

Regulations

Get to know the regulations governing pesticide application. The most contentious form of pesticide spraying is that carried out by aircraft. The rules governing aerial spraying are often broken. Below are some of the most important rules which spray pilots must comply with.

Rules of aerial spraying

Before spraying the company should have informed

- the police;
- all farmers, growers, land owners and residents using adjoining land, by postcard, giving details of intended spraying;
- the Nature Conservancy Council if land is adjoining National Nature Reserves or Sites of Special Scientific Interest;
- the reporting point for local beekeepers' spray warning scheme or individual beekeepers where no scheme exists.

During spraying the pilot should not normally:

- pass over occupied buildings below an altitude of 200 ft (61 m);
- fly nearer than 200 ft horizontally to a dwelling house, hospital, playground, school or livestock building;
- fly below 250 ft (76 m) over motorways or below 100 ft (30 m) over main roads;
- spray agrochemicals when wind speeds exceed 10 knots.

If you see a pilot doing any of these, report it to the police and the Civil Aviation Authority (CAA). They are both supposed to act on your behalf. If they don't, complain to the chief constable and/or your MP. The full set of rules are available from the CAA publications division as listed on page 171 and are called *The Aerial Application Certificate*, code CAP 214.

Don't despair!

The previous notes might suggest that it is very difficult to claim against pesticide damage. Well it is in some cases, if you have an obstinate sprayer, unhelpful police or some other obstacle. But more and more people are complaining effectively, and as they do so, the public and official awareness of pesticide problems continues to grow. If you have a problem, make a fuss!

● *What the government could do*

The gross inefficiencies of current pesticide use, and the many ways in which users can become contaminated, mean that this is an area where improvements could easily be made with quite minor changes in legislation and a relatively small research budget. Amongst the more urgent requirements are:

- Obligatory training for *all* commercial pesticide operators, regardless of how long they have been using pesticides. Training courses should include details of environmental effects and threats to other people, as well as personal health risks.

- Tighter regulations on the provision and use of protective clothing, along with additional research on its effectiveness.

- Introduction of closed systems for the mixing of pesticides, and filling spray machinery, to minimise user contamination at this particularly dangerous stage of the operation. A date should be set for this to become a legal necessity.

- A ban on existing wide spray spectrum nozzles, phased over a number of years, and introduction of safer CDA systems. Standards for spray nozzles should insist on a safe spectrum of droplet size production for all applications.

- Labelling chemicals with recommended droplet size and recommending application rates in droplets per square centimetre rather than gallons per acre, thus allowing more effective use of the new, sophisticated, spraying machinery.

- The obligatory addition of drift-inhibiting vegetable oils to chemicals sprayed in a water medium.

- A ban on ester formulation herbicides to reduce vapour drift.

- A ban on aerial spraying on arable land, where adequate alternatives exist. Spraying of bracken and forestry should include obligatory use of ground markers and compulsory notification of all people in the area about time of spraying and substances used.

- In addition, further restrictions on substances cleared for aerial spraying are also needed, including the removal of more hazardous chemicals, such as benomyl, captafol and 2,4,5-T amongst others.

- A legal obligation to keep public footpaths clear of spray, so as not to contaminate members of the public, and compulsory marking of footpaths when spraying is taking place on adjoining land.

- A legal obligation on sprayers to inform neighbouring beekeepers of their plans to use pesticides, rather than the voluntary schemes which, at present, rely on the beekeepers to take the initiative.

- A central compensation fund for spray drift victims, funded from a direct levy on agrochemical manufacturers, to avoid the very long delays in obtaining compensation which can currently ruin affected growers, or injured members of the public.

- Additional research on the movements and concentrations of low level pesticide contamination.

chapter three

Pesticides in our food

In 1983, the British Association of Public Analysts dropped a quiet bombshell with their report *Surveys of Pesticide Residues in Food*. They found that more than a third of the vegetables and fruit they sampled contained detectable amounts of pesticide residue. Some of these agrochemicals were already banned, or very severely restricted, in the UK (including DDT) while others turned up in food for which they had no clearance. Many were present at levels above the official reporting limits. Yet the government responded blithely by saying that 'in most circumstances, occasional exposure to higher than average levels of pesticide in foodstuff has no public health significance'.

• Pesticide residues in the British diet

The full details of the Analysts' report make alarming reading. They were brought to public attention by a Friends of the Earth press release in 1984, launching FOE's pesticide-free food campaign, which used glossy posters to put across the message about pesticide residues in food. The facts were then reported widely in the media, including a long article in the *Observer*.

Out of 32 lettuce samples, 13 showed contamination, including 'reportable' levels of **lindane**. Lindane is a highly persistent organochlorine pesticide which has been identified as a potential carcinogen and teratogen. It has been banned or severely restricted in at least 15 countries, including Argentina, Hungary and New Zealand. Three other samples contained residues of DDT, despite its use on lettuce having been withdrawn many years previously. A third of the cucumber samples also contained lindane. A third of the tomato samples tested were contaminated with pesticides, including some containing **aldrin**. Aldrin has been banned altogether in Hungary and the USSR and severely restricted in many other countries. It is not permitted for use on tomatoes in Britain or sold for home garden use. Aldrin is thought to cause birth defects and is a suspected carcinogen.

Fruit was also contaminated. A fifth of the apples examined had detectable pesticide residues, including five examples of DDT contamination, despite the fact that DDT had not been recommended for use on apples for many years. Cherries contained low levels of a banned form of lindane. A full breakdown of the results of the survey is given in the table below, and vegetables shown in the table opposite. The table on page 60 identifies the pesticides used.

Table of pesticide contamination in fruit

Fruit	Number sampled	Number contaminated	Number of different pesticides present
Apples	42	9	5
Blackcurrants	11	4	2
Cherries	11	5	6
Gooseberries	11	5	4
Grapefruit	13	2	2
Grapes	23	6	6
Lemons	12	3	3
Oranges	20	8	5
Pears	37	19	5
Plums	15	4	2
Raspberries	17	10	5
Rhubarb	3	3	2
Strawberries	35	14	17
Tomatoes	33	11	6

No residues were found on limited samples of apricots, bananas, dates, figs, kiwi fruit, mango, melon, peaches, pineapple or red currants.

Table of pesticide contamination in vegetables

Vegetable	Number sampled	Number contaminated	Number of different pesticides present
Artichoke	1	1	1
Aubergines	3	2	2
Beans	4	1	1
Beetroot	1	1	1
Cabbage	17	2	3
Carrot	4	2	2
Cauliflower	13	3	4
Celery	4	2	2
Courgettes	5	4	3
Cucumber	16	6	2
Lettuce	32	13	7
Mushrooms	39	12	8
Onions	3	2	1
Peppers	5	1	2
Potatoes	5	5	2
Spring onions	3	1	2
Turnips	2	1	1
Watercress	3	2	3

No residues were found on limited samples of broccoli, brussels sprouts, chicory, chillies, coriander, fennel, korella, leeks, marrow, parsnips, patra, peas, radishes, spinach, sweetcorn or wheat.

Of 305 fruits tested 31 showed residues at or above reporting limits (10.2%) and another 72 showed residues within the limits of detection (23.6%), meaning that 33.8% (i.e. just over a third) had detectable pesticide residues.

Of 178 vegetable samples, 37 showed residues at or above reporting limits (20.8%) and another 24 had residues within the limits of detection (13.5%) making 34.3% (again just over a third) with detectable pesticide residues.

These figures become even more significant if the major British crops are examined.

Major British fruits (apples, blackcurrants, cherries, gooseberries, peaches, pears, plums, raspberries, strawberries and tomatoes) examined showed 37.8% containing detectable pesticide residues.

Major British vegetable crops (beans, beetroot, broccoli, cabbage, carrot, cauliflower, cucumber, lettuce, mushrooms, onions, parsnips, peas, potatoes and turnips) examined showed 35.3% containing detectable pesticide residues.

Pesticides present on food were as follows (Survey I):

Pesticide	Food
Aldrin	spring onions, mushrooms, tomatoes, watercress
Biphenyl 19	lemons
Biphenyl 33	oranges
Captan	cabbages, cauliflower
DDT/DDE	blackcurrant, strawberries, lettuce
Demephion	spring onions, cherries, tomatoes, lettuce, strawberries, mushrooms, cauliflower
Diazinon	raspberries
Dichlofluanid	blackcurrants
Dieldrin	courgettes
Dimethoate	grapefruit, cherries, gooseberries, strawberries, plums, pears, mushrooms
Dithiocarbamates	raspberries, beetroots, lettuce, potatoes
Fenchlorvos	strawberries
Fenitrothion	raspberries
Fenpropathrin	apples
Fenvalerate	peaches
Heptenophos	apples, mushrooms
Hydroxybiphenyl	grapefruit, pears, oranges
alpha HCH	watercress, apples, cherries, gooseberries, raspberries, rhubarb, tomatoes, pears, cauliflower, aubergines, onions, cucumber, cabbage, courgettes
beta HCH	courgettes, mushrooms, watercress, apples, pears, celery, peppers
gamma HCH	tomatoes, artichokes, aubergines, apples, cherries, grapes, rhubarb, cucumber, lettuce, celery, cabbage
delta HCH	cherries
lambda HCH	turnip
Malathion	grapes, tomatoes, lemons
Parathion	mushrooms
Mevinphos	cherries, strawberries
Organophosphorus (unknown)	lemons
PCNB	grapes
Permethrin	pears
Pyrimiphos methyl	mushrooms, strawberries
Tecnazene	potatoes
Thiabendazole	grapes
Thiocarbamates	gooseberries

The Friends of the Earth publicity generated a flurry of alarm. People who had been deliberately cutting down on meat to avoid high levels of hormones and antibiotics now found themselves facing the prospect of ingesting dangerous pesticides with their fresh fruit and vegetables instead. Despite the government's claims that residues were way below the danger level, many people stayed worried. The memories of other 'safe' chemicals, which had subsequently been banned, gave little cause for confidence.

● *How do pesticides get into our food?*

High pesticide residues find their way into our food in a number of ways. Carelessness or ignorance on the part of the grower is often a factor. Too much chemical is applied, or the wrong kind of chemical, or crops are sprayed too near to the harvest. Some pesticides are extremely persistent in the soil, and can be picked up by crops planted in succeeding years. A few persistent chemicals, such as DDT, have been banned in many countries, but others remain on the market. Since DDT and many other organochlorine chemicals have been banned in Britain, their residue levels in food have fallen. Unfortunately, others appear to have taken their place.

> Sometimes a known pesticide can have greater effects than expected. In 1985, the United States government ordered the destruction of ten million watermelons in California after at least 108 people had been poisoned by **aldicarb**, suffering vomiting, diarrhoea and blurred vision. Seventy cases were confirmed in California and a further 38 in Washington, Oregon, Alaska and British Columbia, with at least 200 more suspected.

These factors represent unintentional overuse of pesticides. They could be reduced by better training of farmers and growers, improvements in pesticides (including the banning of several now commonly available) and some harsher penalties for misuse. Other reasons for the occurrence of pesticide residues are bound up more with the way in which we, as consumers, expect food to be presented to us, rather than by technical or operational shortcomings.

• Our own expectations about food

Today, we expect food to have a perfect 'appearance', including size, shape, colouring and absence of blemishes, as well as wholesome nutritional qualities. (In fact, appearance is all too often given greater value than nutrition in selection and marketing of food.) Lettuces with a few holes in their leaves will produce a string of customer complaints if they are offered for sale at the local supermarket. Yet in order to avoid caterpillars eating holes in lettuces, the grower often has to spray them with pesticides right up to the minimum time interval before harvesting, and sometimes even beyond this minimum in practice. Customer expectations cause farmers to put increasing reliance on 'cosmetic sprays', which do nothing at all for the food value. It is these late use sprays which are especially likely to be left on fruit and vegetables.

In addition, an increasing number of farmers are contracted directly by major supermarkets or their intermediaries to produce food. These people are understandably desperate to maintain their standards of appearance and keep the contracts they need to survive. Some contractors actually specify which pesticides to use, but there is little control to see that these guidelines are followed in practice, even if they are adequate. The pressures for keeping up appearances are stronger than the pressures for maintaining quality.

• Modern systems of management

Systems of management are also at fault. Whereas in the past a small farmer would have automatically rotated the crops year by year to minimise the build up of pest infestations, this is no longer practised to anything like the same extent. In the rich loam soils of south and east England the same crops are put down year after year. Large, impersonally managed contract farms are again much to blame for this. The managers simply do not have the same long-term interest in the survival of the land as a small farmer-owner.

When you see huge potatoes for sale in supermarkets, the chances are that they will have been produced by similarly huge applications of fertiliser and pesticide. Some potatoes examined by a Ministry of Agriculture Working Party in the early 1980s were found to have levels of the sprout suppressing agent **tecnazene** of 218 times the maximum residue level allowed by the EEC. Tecnazene dusts and granules are meant to have at least a six weeks' harvest interval between application and harvesting of potatoes.

> Another pesticide which might turn up on potatoes is Temick, a systemic insecticide with the active ingredient **aldicarb**. Temik, manufactured by Union Carbide (the multinational owning the Bhopal factory in India) is an extremely toxic chemical which has caused problems of groundwater contamination in Florida in the USA and other places. It is fatal if swallowed in quite small quantities and can also prove fatal if absorbed through the skin or eye. A report in the November edition of *Farming News* also links Temik with AIDS. Dr. Leon John Olsen and Dr. Ronald Hindshill, both working in the United States, believe that minute quantities of aldicarb can damage the immune defence system in animals and humans leaving them more vulnerable to AIDS and other diseases. The scientists say that Union Carbide has 'engaged in a campaign to frighten the editors of noted scientific publications so that they won't publish our findings'. They are particularly worried about the effects of aldicarb in drinking water on pregnant women, whose immune systems are already suppressed to allow the foetus to develop.

The British authorities explain away residues in potatoes by confidently stating that these are lost during preparation, because they are concentrated in the skin. But what about baked potatoes? And all the people who leave the skins on because this is where many of the nutrients are found? Well, apparently cooking destroys most of the residues present. Despite these bland assurances, we still do not know very much about the way in which residues are passed on in the diet. We don't know what cooking in a microwave does to pesticide residues. Changing farm management systems are putting the onus on the consumer to deal with possible residues, rather than on the grower to avoid them in the first place.

It is not only purchased food which can be contaminated. In June 1984, Cumbria police issued a warning to
people picking wild mushrooms that fields in the
county were being sprayed with potentially hazardous
pesticides. The *West Cumberland Times and Star* reported that the alert followed an incident when a man
who had been picking mushrooms in a field near Friz-
ington was found suffering nausea and sickness. A
spokesman for the Cumbria police said: 'Care should
be taken when choosing sites for picking. The spray
used is not lethal but can cause irritation, nausea and
sickness.'

Storing more food, more of the time

Food storage has always been an essential part of farming.
Another cause of pesticide residue problems today is the way
in which food is stored in modern farm systems. The growing
mountains of agricultural surpluses, and modern consumer
expectations for food to be available out of season, means that
vegetables and fruit are stored for long periods and pesticides
are frequently used as a way of preventing deterioration by
pest attack or fungus.

Almost all British apples, for example, will be sprayed
with a coating of fungicide, often **benomyl**, to avoid
'one rotten apple contaminating the whole barrel' as
the old catchphrase would have it. This coating will
probably remain on the fruit when it is sold in the high
street shops. Yet benomyl is a suspected carcinogen,
and may cause birth defects and genetic damage. It is
irritating to the eyes and skin and has been partially
banned in some other countries, including Finland and
Sweden. However, many other chemicals are also
sprayed on apples, and the government's '*Approved
Products' handbook* lists an amazing 61 different formulations as being useful for controlling one or more
pests and diseases of apples.

Our 'requirements' for year round supplies of all kinds of fresh
fruit and vegetables did not just happen by accident. The food

industry has consistently promoted various types of fresh food with a particularly pristine appearance, which require high use of pesticides and fertilizers to grow and store. Chemical companies have developed agrochemicals to fit the bill, and do very well from the sales. Like the chicken and the egg, it is sometimes difficult to see which came first. But now consumers are 'hooked' onto regular supplies of beautiful looking foods, 365 days of the year. The increased risk of high pesticide residues is probably an inevitable side effect of this process.

There have been some recent changes in the law in Britain to give Parliament the right to set upper limits for pesticide residues. However, at present it is still not clear how these rules will be used, what levels will be set and how food will be tested. Meanwhile, virtually no research is carried out into the low level effects of constant exposure to pesticide residues.

● *The Third World*

In 1971/72, wheat and barley seed treated with **methyl mercury** fungicide was imported into Iraq. This treatment is highly poisonous and should only ever be used on grain for planting, and not for human consumption. Warnings on the sacks were printed in English and Spanish, but not in Arabic. Many of the farmers did not read at all, and few were fluent in a foreign language. It was a time of food shortages in Iraq, which would inevitably tempt hungry people to eat the grain now rather than wait for a new crop to grow. In any case, some of the grain arrived too late for planting. Faced with immediate hunger, many farmers and their families ate the treated grain and fed it to their livestock. Mercury is one of the most poisonous substances ever used as a pesticide, and has been virtually banned throughout the EEC in any form (although it is still used on turf in Britain). In addition to direct toxic effects, it can cause long-term damage to the brain. Huge numbers of people became ill, and many died. Official Iraqi figures put the eventual death toll at 456, but journalist Edward Hughes conducted an on-site investigation and concluded that deaths were likely to

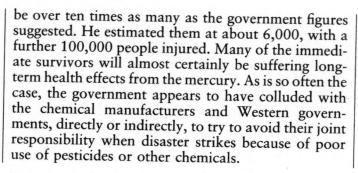

be over ten times as many as the government figures suggested. He estimated them at about 6,000, with a further 100,000 people injured. Many of the immediate survivors will almost certainly be suffering long-term health effects from the mercury. As is so often the case, the government appears to have colluded with the chemical manufacturers and Western governments, directly or indirectly, to try to avoid their joint responsibility when disaster strikes because of poor use of pesticides or other chemicals.

Any problems which the rich Northern countries face from pesticide residues in food are multiplied many times over in the Third World. In addition to the factors discussed above (many of which are likely to be more acute in the South), Third World dwellers have a number of pesticide residue issues of their own to cause them trouble. Pesticide is sometimes used illegally for totally inappropriate purposes, such as killing fish to eat. As described above, treated seed has been inadvertently used as food. The wrong pesticides are frequently applied to crops, and harvest intervals are ignored. And the general shortage of food and water in most poor countries (which may be a political shortage rather than an actual shortfall in supplies) means that contaminated food is more likely to be eaten. People facing acute malnourishment do not have the luxury of worrying about long-term health hazards of food contaminants, but they may be affected by these nonetheless.

David Bull, previously of Oxfam's Public Affairs Unit, exposed several particularly bad pesticide residue problems in the South in his book *A Growing Problem* published by Oxfam. One arresting photo from the book shows old pesticide drums being used to store locally brewed beer in Awassa, Ethiopia. In Ghana, it is common practice to kill fish by adding pesticides to water, because the peasants cannot afford nets and pesticide poisoning is 'cheaper'. 'People in fishing villages complain of blurred vision, dizziness and vertigo, but none makes the connection between these symptoms and the poison' according to a USAID development worker.

Other examples turn up wherever investigators dig deep enough. Contamination of wheat with **parathion**, a highly toxic organophosphate insecticide, (including acute inhalation and dermal toxicity) led to 100 or more known deaths in the south-west Indian state of Kerala in 1958. Old pesticide drums have been seen and photographed collecting rainwater for drinking in El Salvador. It is apparently common for crops to be sprayed the day before marketing in Sri Lanka. Examples like this are just the tip of the iceberg of pesticide residue problems in the Third World.

> The use of pesticides is now particularly advanced in India, famous as the cradle of the 'Green Revolution' (of which more in the chapter on the Third World). Permissible pesticide residue limits were also set impressively early in the subcontinent, in 1954, under the regulations of the Food Adulteration Act. Unfortunately, there are simply not the resources or personnel to enforce these levels in practice. Staff corruption is probably also a problem in some areas, and illegal pesticide users can sometimes bribe their way out of trouble. All 300 samples of leafy vegetables taken from the market at Mysore, a small town on the northern end of the Western Ghat mountain range, contained excessive amounts of **BHC**. BHC, also known as **hexachlor**, is strongly suspected of being a carcinogen and was voluntarily withdrawn in the USA in 1978, after failing to meet Environmental Protection Agency requirements. Most of the milk marketed in the Punjab was found to contain DDT levels above the maximum permitted. Over 35 per cent of food from Hyderabad market, in central India, contained illegally high levels of organochlorines. As long ago as 1965, DDT ingestion levels were found to be 20-40 times higher than in British people. DDT has not yet been banned in India so contamination of the population by DDT is still likely to be serious.

● *The circle of poison*

High pesticide residue levels in the Third World create a lot of problems for the people who live there. However, they also create problems for you and me, living in the rich countries, as well. A lot of the most intensive agriculture practised in the developing countries, where pesticide use will be at a maximum, produces crops for export to the North. These crops including bananas, cocoa, coffee, tea and tobacco, are often very heavily sprayed. Governments desperate for foreign capital encourage maximisation of production of foods which can be sold abroad. When we sell hazardous chemicals to the South, a proportion of them come straight back to us in the food which we import. This ironical state of affairs has been nicknamed the 'pesticide boomerang' or the 'circle of poison'.

The example of the USA

Research into pesticide residues in imported food has been much more thorough in the USA than in Britain. Although the US Food and Drug Administration (FDA) only checked for 30 per cent of possible hazardous chemicals in its import screening procedure, it still estimated that ten per cent of food brought into the United States contained illegally high levels of pesticides. The real figures for all pesticides are almost certainly higher. Pesticide residue levels in imported food became so controversial that beef from Mexico, Guatemala and El Salvador was temporarily banned. Analysis of beef from Guatemala and El Salvador showed that 12.6 per cent had DDT levels above USA tolerance limits. Some samples of milk from Guatemala had DDT levels 90 times higher than US limits. Almost half the green coffee beans imported into the USA contain pesticides banned there. Over 15 per cent of the beans and 13 per cent of the peppers imported from Mexico violated FDA residue standards. Some peppers from a shipment sold onto supermarkets had 29 times more residue than allowed under US law.

Authors David Weir and Mark Schapiro of the Center for Investigative Reporting in Oakland, California, disclosed the details about the American pesticides boomerang in their book *Circle of Poison*. They concluded: 'pesticide pollution does not respect national boundaries. As one of the world's largest food importers, we in the United States are not escaping hazardous chemicals simply by banning them at home.' The book contained statistics from the General Accounting Office of the US Congress which suggested that out of ten types of food imported into the USA, an average of 27 per cent had been sprayed with pesticide for which *any residue at all* would be illegal in the USA.

The pesticide boomerang in Britain

The official penchant for secrecy makes it far more difficult to work out what pesticides are being exported from Britain, and testing of imports is not so thorough, so that the impact of the pesticide boomerang in the UK is difficult to assess. The Public Analysts' report quoted at the beginning of this chapter suggests (on the basis of very limited sampling) that imported foods have less residues than those from Britain. However, wheat imported into Britain is known to have on average higher pesticide residues than British-grown wheat. In addition, testing of imported food is woefully inadequate in Britain and we may simply not know what is coming into the country.

A report, *Pesticide Residues and Food: the case for real control*, by Peter Snell and Kirsty Nicol of the London Food Commission, published in 1985, concluded that improving tests on imported food is an urgent requirement for reducing pesticide residues. Even in the USA, where detection is far better, there are at least 130 pesticides used in exporting countries which escape the net of the FDA's two commonest multi-residue tests. The residue issue is unlikely to go away for quite a while yet.

● *What you can do*

Given that we know high pesticide residue levels often exist in food, many people want to take some positive steps to cut down their exposure. There are a number of ways you can go about this:

1. Organic food

The simplest way of minimising exposure to pesticide residues is by avoiding food grown with pesticides. (There are almost bound to be some residues of very persistent pesticides, but this cuts out the majority.) Either grow your own organically or buy the organic food which is increasingly available throughout northern Europe and North America.

Unfortunately, there are several different 'standards' of organic food in Britain, Europe and North America, which makes things unnecessarily confusing; hopefully these will be standardised in the near future. The largest British classification at present is that devised by the Organic Standards Committee, described in detail in the chapter on alternatives. Farmers who wish to sell their food under the 'organic symbol of quality' have to be inspected by trained personnel and thereafter keep to a fairly narrow set of specifications. These include the elimination of all but a very small range of plant-based pesticides. (The symbol is removed temporarily, for example, if an organic farmer suffers spray drift damage.) Policing is not very strict, due to lack of resources, so that some non-organic food probably does get sold as organic because of the higher price this affords. But most produce sold through the main distribution networks will be chemical-free, or at least as chemical free as it is possible to get in the polluted North. The organic symbol of quality, which should be shown on the packaging of all organic produce, is illustrated below:

Organic food is becoming increasingly plentiful and easy to find. It is stocked by Tesco and Safeway and many small wholefood shops, market stalls and co-operatives. However, supplies are still limited and it is more expensive than ordinary food. Demand already outstrips supply to some extent and not everyone can have organic vegetables. It also means that only the better-off consumer can avoid pesticides in this way.

2. Trying to judge whether something has high pesticide residues or not

Once you buy non-organic food there is no way of knowing

whether the potato or cabbage you hold in your hand has high pesticide residues or not. Analytical tests are very expensive, and anyway there is little chance of tracking down all the hundreds of pesticides available. On the whole, very large roots and vegetables are quite likely to have been sprayed. Most of the huge potatoes now seen in supermarkets will have come from intensive systems described earlier. But a good organic grower will be able to produce big potatoes as well, so the customer shouldn't expect old and grotty looking produce necessarily to be healthier either!

3. Pesticides on the plant's surface

In practice, it is safest to regard all non-organic fruit and vegetables as guilty unless proven innocent, and to act accordingly. A large proportion of the pesticide is likely to be stored in the skin of root crops like carrots and turnips, and fruit such as apples. Other crops will have a surface layer on the skin itself, but not always incorporated into the tissues.

Some of this can be treated by washing, which should always be done very thoroughly, even when produce is bought looking beautifully clean. (In fact, especially when crops look beautifully, and suspiciously, clean. . .) Vegetables with a strong skin can be scrubbed hard with a stiff brush.

Many people also believe that it is worth peeling these vegetables as well. As there are also usually extra nutrients in the skin you will have a trade-off between losing some of the goodness and avoiding pesticides.

4. Cooking vegetables

In high-risk crops, cooking should be thorough, especially if the skin is left on as in baked potatoes. It is claimed that high pesticide levels are destroyed by baking or boiling. The effects of microwave cooking are not really known so you may wish to avoid this if possible until more is found out. If root crops like potatoes or carrots are boiled, the stock may build up pesticides, and should not be used in the traditional ways for soup or gravy.

5. Imported food

Particular care needs to be taken with imported foods. It is probably no longer a good idea to use orange or lemon peel in marmalade because of pesticide residues in the skin. Grapes in vineyards are also often heavily sprayed and a growing number of people are becoming allergic to some wines for these reasons. If you get very sick after drinking a particular batch (and you haven't had too much!) avoid this type, as you may be experiencing an allergic reaction.

6. Asking for pesticide-free food

It is rumoured that some of the large food chains test food for pesticides. Ask if your local supermarket, or their parent company, do so. (It is always worth asking because if enough people show an interest, shops will start to take more care of the health aspects of food; this is already occurring to some extent.)

These precautions are all too inadequate. Proper controls on food residues are needed. In Britain, it looks as if the government is finally moving towards this, and so things may become clearer, and gradually better, in the future.

Many of these things are a trade-off between avoiding pesticides and getting the maximum value from food, such as in the skin or by eating raw vegetables. This has to be a matter for personal decision.

● *What the government could do*

The Food and Environment Protection Act of 1985 opened the way for setting maximum levels of pesticide residues for food in Britain. However, there are still very major gaps in both research and necessary legislation about food residues. Changes needed include:

- Food grown in systems that make build up of residues likely should carry a health warning, giving advice on washing, cooking times to detoxify pesticides (if this is possible), etc.

- Maximum limits for pesticide residues in food should be set in line with those in operation in other countries, including the USA. Furthermore, food residue limits should be set at zero for all hazards with no known lower activity threshold (e.g. long-term hazards like carcinogens or teratogens) and for pesticides which are likely to accumulate in the body.

- Food sampling should be increased, and organised in such a way that it is possible to trace the source of contaminated samples.

- Particular care should be taken to increase sampling of food imported from abroad, especially in the cases of countries known to have lax standards of pesticide safety. Food with an unacceptably high pesticide content should be returned, and further imports suspended until the situation improves.

chapter four

Pesticides in the home and garden

In September 1984, a Wimbledon householder accidentally breathed in fumes from a nettle weedkiller containing 2,4-D and 2,4,5-T, while she was mixing it up in a watering can. In a letter to Friends of the Earth she claimed 'I have been feeling very unwell since then. . . . The doctors I have consulted have not been very helpful.' She later described her symptoms as 'hot and cold sensations all over, inflamed sinuses, burning chest and back, my throat has enlarged with the feeling of a lump in it, swellings have appeared. . .'

The story has a familiar ring. It could well be another extreme allergic reaction of the type described in earlier chapters. Or it could be something else entirely unconnected with pesticides. Like the health problems described before, it is usually impossible to 'prove' that an illness is linked to a particular chemical or spraying incident, and each case has to be judged on its own merits.

However, there is something new here. Instead of the sufferer being the victim of someone else's spraying, here she was responsible for breathing in the fumes, while mixing up pesticide for use in her own garden. This was not a heavy duty spray meant to give protection to a huge area of food crop, but something supposed to be used around the house.

Every year in Britain, we spend about £15 million on pesticides to spray on the flowers, vegetables and lawns in our back gardens. We choose from a bewildering range of some 600 or more pesticide brands, made up of over 100 different chemicals. The industry's well known secrecy means that precise sales figures are not available but, as near as anyone can judge, about a kilo of pesticide active ingredient is applied to every acre of British garden, every year, and use is still increasing fast.

Whatever worries we may have about pesticides on the farm, we generally assume that domestic pesticides are somehow inherently safer, more carefully vetted, and sold in such low concentrations as to be virtually harmless. This message has been pushed vigorously by the agrochemical industry. Sales literature and advertisements highlight the attractive environments created by weedkillers and insecticides, whilst minimising the dangers associated with their use. Over the last few decades, garden chemicals have become such an accepted part of the British home garden scene that many people assume that they pose no risk at all.

This is not the case. Many garden chemicals are also dangerous to our own health, to the lives of our pets and to the survival of any wildlife we attract to the garden. Some can cause short-term problems, such as irritation to the eyes, skin and breathing, while others are known or suspected of resulting in long-term and very serious side-effects including cancer and birth abnormalities. Recent research has shown that chemicals openly for sale in garden shops include over 50 which are irritating to eyes, skin and/or breathing, 25 known or suspected **carcinogens** (cancer producers) and 29 known or suspected **mutagens** or **teratogens** (causing mutations and birth defects). Many can kill wildlife; 46 garden chemicals are hazardous to fish for example, while others kill bees and butterflies, or larger animals like pets, birds and hedgehogs.

The continued use of **mercurous chloride** in the garden is a good case in point. Mercurous chloride or 'Calomel' is a fungicide traditionally used on ornamental plants and turf. It is an extremely toxic poison which is fatal in quite small doses and can also cause permanent damage to the nervous system according to the United Nations International Labour Organisation. It is poisonous to all forms of wildlife and the World Health Organisation classify it as tremely hazardous'. Mercury compounds have been banned in many countries, including Chile, Cyprus, West Germany and New Zealand, and the EEC have restricted mercury with legislation banning its use on turf. Yet Britain not only still allows sales of mercurous chloride for use in gardens, in the form of products from Pan Brittanic Industries and ICI, but they are even listed in the trade's own *Directory of Garden Chemicals*, meant to provide sound advice to home gardeners.

The use of garden chemicals poses rather different threats to health and wildlife than those already described for farm and municipal applications.

Garden pesticides are used far closer to where people live, allowing, for example, young children to breathe in pesticide spray or to brush against treated plants in shorts and skirts, thus getting the chemical directly on to their skins.

Different food crops in a vegetable garden are generally planted closer together than on a farm, adding to the risks that one crop will be contaminated with spray meant for another.

Wildlife is often encouraged into the garden, through the use of bird tables and by planting butterfly bushes, only to be killed by thoughtless use of chemicals.

The general British ignorance about chemical hazards is often multiplied many times in garden users who will apply spray in bad conditions, with no protection and with scant regard for where any spillage or excess ends up.

On the other hand, there has been a tacit acceptance of the additional dangers posed by pesticides used in very close proximity to where people live, and a gradual phasing out of some (although not all) of the most intensely poisonous, often as a result of pressure from environmental groups. This, in itself, shows that the claims of complete safety should not always be taken at face value.

Garden chemicals are also, as the industry is at pains to point out, generally used in lower concentrations than those on the farm. But then we have already seen how types of chemical once thought 'safe' have subsequently been re-evaluated and their use withdrawn. Could garden chemicals, then pose real threats to the health of Britain's gardeners?

Garden chemicals and human health

Until 1985, there was very little written about the herbicide **ioxynil**, sold either alone or in various chemical mixtures by such established firms as ICI, Fisons, May and Baker, Pan Brittanic Industries and Synchemicals. The government's own *Approved Products Handbook* simply warned to beware of spray drift, gave some suggested harvest intervals and mentioned damage to fish, all very common points. Then, on 13th June 1985, the Ministry of Agriculture called for a voluntary freeze on retail supplies, after evidence that ioxynil caused birth defects in laboratory animals. Despite the publicity, ioxynil can still be found for sale in many garden shops and centres at the moment. It is also available for use on farms. Friends of the Earth provided a detailed critique of the ioxynil issue to the House of Commons Agriculture Committee.

We have already established that there is no such thing as a 'safe' pesticide. All pesticides are poisons of one kind or another, and many will have harmful effects on our health. The difficulty lies in assessing whether the risks are unacceptably high in the light of our current knowledge about the chemical. This is just as true for garden chemicals as it is for the sprays pumped out of huge booms towed by tractors on arable farms; in fact in many cases the chemicals will be exactly the same.

One of the insecticides which has come into common use over the last few years in domestic gardens is **permethrin**, which has the reputation of being very non-toxic. However, a storm of controversy has surrounded its use in the USA. Dr. Adrian Gross, an American government pathologist of twenty years standing was demoted after writing a 48-page memo accusing his superiors of aiding two medical companies in their efforts to register it. 'Sure', he says, 'permethrin is a carcinogen – I don't know of another insecticide where the evidence is so overwhelming.' Yet it has received clearance in many countries, including Britain.

Commonly used garden chemicals with strongly suspected links with cancer or genetic effects include amintriazole, **benomyl, captan, carbendazim, dichlorvos, maneb, nicotine** and **zineb**. Amintriazole was banned altogether in Sweden after evidence was found of increased cancer in railway workers using it for track clearance. Benomyl has been banned in Finland and other countries because of fears of genetic risks. Dichlorvos, an organophosphorus pesticide, has been identified as a potential carcinogen and mutagen by the US NIOSH, while James Scharden identifies risks with maneb in his book *Chemically Induced Birth Defects*. The highly toxic nicotine (which until recently was actually passed for use by organic growers in Britain) is a known carcinogen but also a suspected mutagen. And work in Russia has pinpointed the mutagenic effects of zineb. There are many other examples listed in the Directory at the back of this book.

Take the herbicide **amintriazole** (also known as amitrole) for example. There is a long record of suspected carcinogenicity with this chemical, including work in the US 'gene-tox' programme, research by NIOSH showing possible birth defects in rats, and statistical research about the incidence of cancer in railway workers in Sweden. Yet British garden chemicals, such as Murphy's and Fison's Path Weedkillers, Murphy's Super Weedex and Agrichem's Hytrol all contain amintriazole without even a warning on the packet of any long term effects, indeed the *Approved Products Handbook* only mentions possible spray drift effects.

Even if chemicals are supposed to be exclusively for farm use, there is little to stop them having a much wider circulation in practice. Friends of the Earth report that on July 19, 1984, an agricultural sale open to the public in Kent contained a number of items containing toxic pesticides. These included 1 lb of **mercury bichloride**, a highly toxic poison, and brand name products which no longer appear in the Pesticide Safety Precautions Scheme list, such as 'Merolan', 'Renardine' and 'Molprox'. Children were walking round the exhibits, and some of the pesticide packets had been opened. There was no restriction on who could bid, although even those which still have clearance for use include a number supposed to be for professionals only. A local FOE member claimed that this kind of unregulated selling was not uncommon.

Hazards to wildlife

Attracting wildlife into gardens is a national British pastime. Millions of homes maintain bird populations through hard winters by regular feeding, provide safe nest sites, put out milk for hedgehogs and choose flowering plants which will attract butterflies and bees. Sadly, many of the same people who take such a pride in their back garden nature reserves are simultaneously killing off a proportion of this wildlife by their use of sprays and pellets.

Despite the wide publicity for insecticides which are 'harmless to bees' or other beneficial insects, the vast majority of garden insecticides kill all insects, irrespective of whether you want them to or not. Careful spraying or placing of slug pellets can go some way towards safeguarding wildlife, but basically if you choose to use pesticides then you will suffer the consequences in terms of fewer colourful butterflies and moths, fewer attractive beetles and, probably, fewer of the larger animals like hedgehogs or toads as well. In addition, it is not generally known that many fungicides also have an insecticidal effect, so that spraying against wilt or moult will also kill insects.

By far the biggest hazard facing larger animals and birds comes from pellets laid down to kill slugs and snails. Chemicals like **metaldehyde** are acknowledged to be a major hazard, both because the animal or bird can eat them directly and, probably more importantly, because animals eating dead slugs and snails will pick up residual poison themselves. Even those people who are careful to hide their pellets under cover precisely in order to avoid poisoning larger animals may well be unwittingly killing them through this more indirect method.

All too many gardeners have had the experience of finding sick hedgehogs or birds in the garden; many of these dying animals will be suffering the effects of direct or indirect pesticide poisoning. However, visible corpses are only likely to be the tip of the iceberg because most poisoned animals will, like sick animals everywhere, crawl away to die in a place where they are unlikely to be found. The effects of pesticides on your garden wildlife will be most obvious by the absence of those animals which were, at one time, a far more common component of the life around the house and home.

Pesticides inside the house

Domestic pesticide use is not confined to the garden of course. We also use pesticides inside houses, probably far more than most of us realise. Pesticides are: applied directly to houseplants; sprayed against flies and mosquitoes; hung up as fly killing strips; put on our pets; left lying around to kill mice and rats and used in various kinds of timber treatment. Some of these pesticides are also very dangerous.

> Workers in a joinery in Hackney, London, felt sick after handling windows and doors treated with a wood preservative called 'Vascol'. They contacted the London Hazards Centre, who contacted Hickson's, the manufacturers of Vascol, to find out the pesticide used. The head office said that the pesticide was **dieldrin**. (Hickson's Barking branch apparently said that the pesticide was **lindane**, but this was later proved to be incorrect.) On March 25 1986 the Barking Health and Safety Executive office put a prohibition order on further use of Vascol-treated domestic joinery in council housing. Robert Sheath, the principal factory inspector involved, told the Hazards Centre that he had been 'flabbergasted' to find dieldrin used in confined spaces in sites with little risk of woodworm attack. 'The balance of need to risk is so low that I couldn't be party to it,' he said.
>
> Rentokil, the pest control agency, had 88 different brand name products in the 1986 Pesticide Safety Precautions Scheme list of pesticides cleared for use in the UK. Many of these will be used inside people's houses. The 25 chemicals used in the products number amongst them lindane (including use against carpet moths) and **carbaryl** (in ant and insect powder). Both of these have a number of suspected long-term health effects and have been banned in several countries.

Again, the argument used is that the small amounts of pesticides involved form, at worst, an insignificant hazard to humans. However, lindane is known to be persistent and to build up in the fatty tissue of the body – the reason it was banned in Hungary. And several products, including DDT, have already been withdrawn for domestic use in Britain following evidence of adverse health effects.

Today, one of the main uses of insecticides and fungicides inside the house is for timber treatment against woodworm, fungal attack, etc. Paradoxically, many of the most hazardous chemicals have, at various times, been cleared for these purposes, including the highly toxic dieldrin.

In September 1984, a wood treatment company applied a number of chemicals to a house in south-west London, including **tributyl tin oxide** and **dieldrin**. The latter has been banned or severely restricted in over 20 countries because of its toxicity and extreme persistence. Clearance was withdrawn in Britain under the Pesticides Safety Precaution Scheme at the end of 1984, although it can still be used as long as stocks last, which means that some will remain on the market for several years yet. It can still be used in paints and lacquers. The World Health Organisation set a recommended upper limit for concentration in water of 0.0003 parts per million.

The fumes from the treatment reached the part of the house where Barbara Kearns lived with her baby son, Jake. The workers did not warn her to beware of fumes at the time and instead left her a note about keeping the place well ventilated after they had gone. She wrote to Friends of the Earth that 'preventative precautions were virtually non-existent'. Both mother and son fell ill, and Jake was admitted to hospital with what turned out to be severe bronchitis. Although both the hospital staff and the Health and Safety Executive were initially quite in agreement that the pesticide fumes could have brought on Jake's condition, they were much less forthcoming when a journalist started asking questions about the case. The company changed their story about what had been sprayed several times and at óne stage said that they had re-used canisters which had contained other pesticides before.

On being discharged from hospital, Jake's blood was sampled and sent for analysis. It was found to contain the following organochlorine pesticides (measured in parts per billion):

    0.50 dieldrin
    2.90 beta HCH
    0.90 DDT
    22.90 DDE (a metabolite of DDT)
    0.50 gamma chlordane
    0.50 heptachlor expoxide
    0.10 transmonachlor
    2.00 hexachlorobenzene

If most or all of the pesticides were present in Jake's blood as a result of his exposure in the house, this would tie in with the firm's statement about containers being re-used. It would, therefore, be impossible to say exactly what would be sprayed in any one day. The case became well known and a question about it asked in the House of Commons. The Health and Safety Executive inspector in the area, Mr. R. Sheath, commented that: 'It does appear that the advice given in the Government Data Sheets, in the Industry's Code of Practice and by most contractors is not sufficiently comprehensive. I am now seeking improvements.'

These, and other improvements will only come about when more people become worried about the pesticides which they use around them in the home, and start telling the government about their concern.

• *What you can do*

Many of the pesticides we use are of very marginal benefit, and can generally be omitted without very serious consequences. However, if you really want to cut down on pesticide use overall in the home and garden, or eliminate it altogether, then you are going to need some other ways of getting rid of the real pests which otherwise cause a nuisance.

The first stage is to minimise the use of the most hazardous pesticides, as outlined in the Appendix and described throughout this book. Then there are also some pesticides which have been cleared as particularly safe, and the chemicals passed for use by the Soil Association can be found on page 166. In addition, over the last few years, organic gardeners have built up an impressive range of pest controls which don't rely on chemical sprays, powders or pellets. Some of these are traditional methods, from the days before pesticides were available, others are new or newly adapted. Some work very well indeed, others work sometimes and a few don't seem to be worth the trouble. (I remember laboriously building a scarecrow to preserve a freshly planted area of grain, only to see the neighbourhood crows using it as a perch within minutes of its erection.) Some methods work well only on a small scale, so that they are suitable for a back garden but impractical for any commercial grower. You will have to experiment.

Below, some of the better known methods of pest control without poisons are briefly explained.

Lawns and worm killers

The only lawns where worm killing pesticides are justified (perhaps) are golf courses and bowling greens. If you find the tiny piles of mud which the worms leave on the surface (their castings) a nuisance, they can easily be swept or raked away. The worms help aerate the lawn by tunnelling through the surface and help maintain the rapid cycling of nutrients which keeps a garden healthy. Indeed, organic gardeners actively encourage worms for just this reason, and a few actually build up 'worm farms' where worms are bred in ideal conditions for later release to soil or compost heap. Dr. Victor Stewart, Emeritus Professor from Aberystwyth University, made an international reputation for 'solving' sward problems on cricket pitches and football fields, which he did by advising them to build up their worm populations again. The alternative way of controlling worms in turf is not to control the worms.

Fruit

Many diseases of fruit can be controlled by care and good management; removing sick-looking shoots, leaves and fruit before pests have time to get a real grip on the plant, pruning at the correct time and so on. The Henry Doubleday Research Association have published a fairly comprehensive guide, 'Pest Control Without Poisons', which gives suggestions for controlling most of the commoner pests without resorting to sprays or powders. The table below summarises a few of these methods. However, I do not attempt to give detailed advice here and suggest that you refer to the books listed in the bibliography for more detailed information.

Pest control on fruit without using pesticides

1. Removal of all wilting or sick looking parts to control pests. Once removed the parts should be burnt. E.g.:
   - sick blossoms of orchard trees to control the pith moth;
   - unnaturally large pear fruit to control the pear midge;
   - unnaturally large buds of currant bushes to control the currant gall mite;
   - obviously maggoty fruit to control the codling moth.

2. Removal of plants which serve as intermediaries or alternative food plants for pests. E.g.:
   - removal of dead nettles and sow thistles from the garden to prevent currant aphids from continuing to survive in the garden.

3. Management techniques and timing. Pruning at the correct time can reduce or eliminate infestations of some pests. E.g.:
   - apple and pear aphids can be controlled by summer pruning.

4. Barriers to prevent pests reaching the fruit. E.g.:
   - grease bands around the base of fruit trees are a traditional and effective method of controlling pests which reach fruit by climbing up the trunk. Used against the winter moth, fruit weevils, etc.

5. Trapping pests. E.g.:
- sacking wrapped around the base of an apple tree at the correct time of year will attract the apple blossom weevil. The sacking can then be removed and burnt.
- newspaper placed on the ground at night will catch fruit weevils, which drop from trees if a strong light is shone into the branches. The paper can then be burnt.
- removing the bracelets of eggs left on apples and pears by the lackey moth.

Vegetables

Vegetables can also be treated by fairly similar methods of using ecological cunning rather than chemical warfare. Again, I have not attempted to provide a detailed guide here, but the list below gives an indication of some of the methods available.

Pest control on vegetables without using pesticides

1. Removal of intermediary plants or alternative food sources, to prevent build up of particular pest species. E.g.:
- shepherds purse, the common weed which is a member of the cabbage family (and used to be a source of food in the Middle Ages) is also host to club root, and so special care should be taken to remove this weed.

2. Time of planting can reduce those pests which only emerge at a certain period of the year. E.g.:
- planting early varieties of bean plant can eliminate blackfly, which does not emerge until later in the summer;
- reducing the growing time of onions, by using sets rather than growing from seed, can help reduce onion fly.

3. Barriers to pests are again used, with both physical barriers and scent barriers (to stop pests locating their food plant) are utilised. E.g.:
- a four inch square of roofing felt around the base of cabbage, pushed through when the plant is very young, will prevent cabbage root fly from crawling down through the soil;
- interplanting onions with carrots will confuse the carrot root fly and reduce the incidence of attack.

4. Trapping is again possible. E.g.:
   - leatherjackets (the larvae of crane flies or 'daddy long legs') can be trapped in lawns because they tend to come to the surface on damp nights. Watering the lawn, then putting down sheets of black plastic sheeting fools the leatherjackets into emerging. After about four hours the sheets can be removed and a lawn mower or roller pulled over the area, with the squashed larvae left as food for the birds.

5. Rotation of crops into different beds each year prevents build up of pests which are specific to one species. E.g.:
   - the cabbage root fly will remain in the soil and attack the plants every year. However a four year rotation (e.g. cabbages, potatoes, carrots and peas) will ensure that the pest has disappeared when the cabbage is next planted on the bed.

Slugs

Slugs are often regarded as the Achilles' heel of organic gardeners. Most of the accepted methods, such as attracting them to saucers of beer, or putting up barriers of ashes or sand, stand little chance of succeeding against a determined attack. But on the other hand, the molluscicides used are amongst the most dangerous chemicals from the point of view of wildlife. The method I have always found best is just to go out on damp nights a few times; say two or three days at the beginning of the season and occasionally thereafter. You will soon learn to pick out slugs fairly efficiently by torchlight, pick them up (in gloves if you're squeamish!) and kill them. The accepted method is dropping them into salt water although I think this is unnecessarily cruel, as it takes them quite a long time to die. A gumboot is much quicker.

Biological control

The other very major way of controlling pests is to actively encourage their predators to take up residence in the garden. And without pesticides, the predators are more likely to survive as well. The control of slugs is a good example. Many creatures, like hedgehogs, toads and some birds, will happily make a meal of your slugs all summer long given half the chance. But these animals have a number of requirements over and above food, such as an area of 'wild' vegetation to live in, a pond (for frogs and toads), and sometimes nestboxes for birds. There are now many excellent books on making wildlife gardens specifically to attract these animals, and some are listed under Books.

● *What the government could do*

Many of the legislative changes outlined in previous chapters also have a bearing on garden use. However, there are some additional factors in the home use of pesticides which need addressing separately, including:

- Banning some of the more obviously dangerous chemicals still used in back gardens, including atrazine, captan, chlordane, dimethoate, gamma HCH, paraquat and 2,4,5-T. This has already occurred in several European countries and companies in Britain could change the formulations of garden brands in line with mainland Europe.

- Making the use of childproof containers mandatory for all garden and domestic pesticides, in the same way as all medicine bottles now have childproof caps.

- Improving labelling of garden pesticides, which is especially confusing because they often come in very small containers.

- Increasing the amount of public education about the potential hazards of garden pesticide products, including posters for display in garden centres, government sponsored advertisements in gardening magazines and inclusion of greater safety details in gardening books, radio and TV shows.

- A thorough overview of the clearance of chemicals for domestic timber treatment is needed.

chapter five

Wildlife

On Easter Sunday 1985, a party of young canoeists were paddling along the River Avon near Evesham, when they found two adult herons floating dead in the water. The corpses reached members of Evesham Friends of the Earth, who immediately sent them to the Institute of Terrestrial Ecology (ITE) for analysis. Publicity in the local media resulted in over 50 telephone calls reporting other dead and missing birds, including more herons. In all 17 dead herons were found, and eight bodies recovered for examination. It became apparent that the birds had started to die in the River Avon early in the spring, following a hard winter. Subsequently, an important heronry has entirely disappeared.

Analysis by ITE and the Hereford and Worcester county analyst showed that the birds had been killed by a lethal mixture of DDE (the breakdown product of DDT), dieldrin and polychlorinated biphenyls, although levels of each individually would have been sufficient to kill many of the birds. Scientists reported that levels of DDE were higher than they had ever seen before. The voluntary ban on DDT had already become law. Some of the pesticides could have been stored in the birds' fat and released in the very cold weather of the previous winter, but this seemed unlikely to account for all the contamination.

Further investigations by FOE discovered stocks of

DDT still illegally for sale in the area. Friends of the Earth members actually bought some DDT before issuing a press release about the sales. These 'banned' chemicals had wiped out a whole colony of important freshwater birds.

The colossal damage that pesticides have done to wildlife is one of the greatest and least recognised of environmental tragedies in the rich countries. Yet threats to wildlife were the first pesticide issue to be voiced by environmentalists, over 25 years ago. Why are things still so bad today?

• A history of concern

For many contemporary environmental activists, the publication of Rachel Carson's book *Silent Spring* in 1962 marked the real beginning of their involvement in conservation issues. Ms. Carson was already an internationally renowned marine biologist, who had published several best-selling books about the sea, when she completed her sustained polemic against pesticide use in the USA. In rich and often emotive language, she blew the lid off what was happening to wildlife in huge areas of the United States:

> '... who has decided – who has the right to decide – for the countless legions of people who were not consulted that the supreme value is a world without insects, even though it be also a sterile world ungraced by the curving wing of a bird in flight?'

The book exploded onto the American scene and changed the attitudes of a generation. Chemical firms attacked her ruthlessly, and often personally, publishing glossy brochures like *Silent Autumn*, claiming that agrochemicals were completely vital to America's livelihood. Pesticide manufacturers poured scorn on her scientific veracity and accused her of political subversion. No matter. No matter either that parts of her book were overstated, that there were a few technical errors. Rachel Carson's book changed a little bit of history and the simmering unease about use of chemicals found a powerful voice. *Silent*

Spring has remained constantly in print ever since, despite being way out of date with respect to today's pesticide effects.

The crux of Ms. Carson's thesis was that pesticides cause damage to wildlife in more subtle ways than straightforward poisoning. One of the most important of these side-effects was the possibility of persistent chemicals (i.e. those remaining unchanged in the bodies of creatures which ingest them) becoming concentrated in the food chain and damaging animals far removed from the place where pesticides were used.

• DDT

Take a very simple example: the passage of DDT up the food chain. DDT, you will remember, was one of the first 'wonder chemicals' and was extensively used throughout the world in the decades following the Second World War. It is extremely persistent, building up in any creatures which ingest it and capable of being stored for long periods in the body fat.

Suppose that DDT is used to kill insect pests in a certain field. A proportion of the insects will not die immediately and will, instead, be eaten by birds and small mammals. These larger creatures will also pick up excess pesticide from the soil and from vegetation. DDT at these low levels may not be fatal to small birds and mammals in the short term, and they carry on life as usual, all the time building up steadily higher levels of DDT. And, as is the way of things, a good number will be eaten by the sparrowhawks and kestrels which patrol the field hunting for prey. A normal hawk will eat several birds or mammals in a week, and hundreds in a lifetime. Each of the smaller creatures will have built up DDT from their own prey, so pesticide accumulated by several thousand insects can all end up in the body of one bird.

The effects of DDT, and other persistent organochlorine pesticides, on birds of prey has been catastrophic. High DDT levels upset the reproductive ability of the birds, making them infertile or liable to produce eggs with very thin shells which break. Populations of some British birds of prey plummeted under the twin onslaughts of pesticides and habitat destruction. In Norfolk, there were several hundred pairs of sparrowhawks in 1949, but by 1965 numbers had fallen until only a single pair was left in the county. Peregrines also fell dramatically in numbers, and large numbers of their eggs did not hatch. In some cases the birds' behaviour also changed so that eggs were not incubated or, in extreme cases, were broken open by the parents. High organochlorine levels were found in both eggs and bird corpses. Golden eagles, Britain's largest bird of prey, also fell in numbers over the post-war period. The role of DDT, and other persistent organochlorine chemicals in this is now no longer open to doubt.

Scientists think that persistent chemicals like DDT have spread virtually throughout the whole world. A study of the Adelie penguins in Antarctica showed that today's birds have detectable levels of DDT (about 0.2 ppm) in their fat whereas an old specimen (found frozen beside Captain Scott's famous base hut) did not contain detectable pesticide residues. There has been much debate as to how the DDT got there, with some people opting for the theory that it had accidentally been discharged from US ships, while others believe that it was carried across in the bodies of plankton which had drifted thousands of miles from more polluted waters. We simply do not know.

However it got there, the evidence from the Antarctic, the Arctic, and other remote parts of the world, suggests that pesticides are now a truly global phenomenon. As Kenneth Mellanby, formerly the Director of Monk's Wood Experimental Station, wrote in his book *Pesticides and Pollution*:

> '. . . pesticides can apparently be transported throughout the world. Pockets of high concentrations seem to develop in unexpected places, depending on meteorological factors, ocean currents and the activities of man.'

The public outcry following the publication of Rachel Carson's book, and a few other important revelations, built up into a powerful lobby demanding a ban on the most dangerous agricultural chemicals. Many countries brought in legislation forbidding manufacture and use of DDT and other pesticides particularly lethal to wildlife. Current legislation owes much to those early campaigns. DDT, for example, is now banned or severely restricted in at least 20 different countries plus the

EEC, and including nations as diverse as Bulgaria, Cyprus, Norway and the USSR. In Britain, a voluntary ban operated for years, but DDT was not finally banned until October 1984 and, as we have seen, it has remained available since that time. Although use had greatly decreased since its heyday in the 1950s and 1960s, there were signs that it was actually increasing again slightly in Britain just before the ban.

Although DDT is still very widely used in some Third World countries, with serious wildlife effects, use of most very persistent organochlorine pesticides has fallen off very sharply in the North. This has been matched by a parallel increase in birds of prey in Britain, with peregrines and eagles recovering numbers particularly well. Importantly, this increase has also been accompanied by the public perception that things have now much improved and that pesticides are now no longer such a danger to wildlife. Unfortunately, this general feeling of complacency, much promoted by the chemical industry, is simply incorrect. Pesticides are, if anything, a worse problem now than they were 20 years ago. (And they are certainly causing far, far more serious effects in many developing countries.) More types of chemical are available, far greater amounts are being used every year and larger areas are being treated than ever before.

• Pesticides today: getting worse rather than better?

Joyce Tait of the Open University identifies a fundamental difference between the official approach towards pesticides and humans and governmental attitudes about pesticides and wildlife effects. Manufacturers, and governments, start to get worried if there is any substantial evidence of damage to humans (although not worried enough as I have argued previously). But they only start to get worried about wildlife if whole species or populations are at risk from pesticides; routine poisoning is accepted as a normal part of any agricultural operation, regrettable perhaps, but unavoidable.

Take the British government's own statements for example. In the now-defunct *Approved Products for Farmers and Growers* book of 1983, some 132 pesticides were identified as harmful to fish, many of these being listed as 'dangerous' or 'extremely dangerous'. Fully 42 per cent of insecticides were listed as harmful to wildlife, along with a number of herbicides and fungicides. Of course, virtually all insecticides are going to harm other insects along with the 'pests', so their use automatically endangers insect types like butterflies and moths. In addition, it is now known that fungicides pose a greater risk to invertebrates than generally acknowledged. And so on. This particular guide only includes the safest chemicals passed for use by the government, and only categorises them according to correct usage.

The actual risks are far higher. Dr. Tait's survey, referred to in the section on human health hazards, also includes a section on whether farmers had noticed dead birds or animals after spraying. Many said that they had. And yet these visible corpses are likely to be just the tip of the iceberg because most creatures will crawl away to somewhere quiet and safe if they feel unwell, so that their bodies are unlikely ever to be discovered. It is only in the last few years that the real risks faced by wild animals and plants have been studied in much detail with reference to modern agricultural spraying systems.

Every year, over 500,000 buildings are subjected to remedial timber treatment in Britain. Most of this treatment is for wood-boring insects, especially woodworm. And most of the chemicals used are highly toxic organochlorines such as **lindane** (also known as gamma HCH) and, in the past, **dieldrin**. Both of these insecticides are highly lethal to bats and are extremely persistent. A building treated 29 years previously was still found to contain lethal levels of lindane when tested recently. As their natural habitats decline, bats increasingly rely on old buildings for roosts in the agricultural areas of Europe and North America. Yet timber treatment can destroy colonies and stop other bats from using the site for literally decades.

Even fungicides like **pentachlorphenol** have been shown to kill pipistrelles, Britain's commonest bat, a year after treatment, acting both by contact and inhalation. This means that they are almost certain to be toxic to the rarer bats as well. Many bats are now extremely uncommon and some are threatened with extinction. All British species receive special protection under the 1981 Wildlife and Countryside Act. The greater horseshoe bat has declined by over 90 per cent during the present century, most of this being in the last 30 years. Apparently the government is now considering whether lindane should be withdrawn because of its effects on bats and dieldrin is no longer cleared for use. Yet lethal chemicals continue to be used, despite the existence of less harmful alternatives.

Today's pesticide problems fall into four main categories:

1. Poisoning of fairly large creatures through misuse (accidental or deliberate) of highly toxic chemicals. Lindane poisoning of bats comes into this category.

2. Poisoning by new agricultural chemicals which have been inadequately tested for harmful effects on wild species.

3. Contamination of large areas of farmland, or neighbouring woods and hedgerows, so that plants and smaller animals are killed, often without the sprayer noticing that this is taking place.

4. Contamination of freshwaters, killing plants, invertebrates and fish.

Each of these will be discussed in some detail below.

Poisoning of larger creatures

Deliberate poisoning of birds of prey, or mammals, is still a major problem in some parts of Britain. It reaches crisis proportions in some areas of Europe and in certain Third World countries.

Because some birds of prey and carrion eaters like crows, peck at dead sheep and lambs, many farmers believe they will also attack healthy sheep, or finish off sick animals. They don't, but this doesn't stop poisoned bait being laid every year for these unfortunate, and innocent birds. Recently, some of the extremely rare red kites were killed in this way in mid Wales, despite the national population being little more than 30 pairs. Buzzards, carrion crows and ravens are all common targets. Farmers sometimes construct baited cages to entice birds into the trap where they are poisoned or shot. I have seen one of these myself, hidden away in a secluded wood in mid Wales, with three or four carrion crows lying dead inside. Foxes are not infrequent targets for pesticide poisoning. Special poison bait is laid for moles to prevent them disturbing the sward of pastureland. Rodenticides are used to kill mice and rats but also affect many other small rodents and other mammals and birds (including pets of course).

Sometimes poisoned 'bait' can be laid unintentionally. Since 1970, many hundreds of wintering wildfowl species, including greylag geese, pink-footed geese, brent geese and Bewick's swans, have been killed in a series of incidents involving them eating seeds treated with **carbophenothion,** an insecticide used to protect seeds against weed bulb fly. Between 1971 and 1974 about 1,600 geese were killed in five incidents in Scotland and northern England, resulting in carbophenothion being withdrawn as a winter seed dressing in east Scotland and Humberside on a voluntary basis. Since then several further incidents have been reported, all in East Anglia, and over 140 birds found dead. This is probably an underestimate of the total numbers involved.

Poisoning by new chemicals

The introduction of new chemicals can sometimes lead to problems, if their toxic effects have not been adequately studied (although tighter regulations should now make this less likely to occur in Britain).

Two poisons introduced into Britain fairly recently have had a disastrous effect on the population of the barn owl, which has declined by 10 per cent in the last few years, with a far greater population collapse in some areas. The two pesticides, **difenacoum** and **brodifacoum,** have been replacing traditional rodent poisons like warfarin in places where rats have become partially resistant to these. Unfortunately, these new pesticides can build up in the bodies of many small mammals and, as with DDT, they accumulate in the tissues of the top predator until they reach lethal levels or affect breeding success. Deaths of pheasants on a Hampshire farm were also caused by brodifacoum, and the Hawk Trust called for its withdrawal. Brodifacoum and difenacoum should now be banned in Britain, because they never had clearance under the Pesticides Safety Precaution Scheme, which is now a legal requirement under the 1985 Food and Environment Protection Act. However, experience with DDT and other 'banned' chemicals shows that the two are likely to remain illegally available in some places.

Even if difenacoum and brodifacoum are eventually eliminated completely, many other legally available pesticides are known to be dangerous to birds. Amongst these hazardous chemicals are carbamates, including **aldicarb** (sold as Temick), which has also been responsible for the deaths of many gulls, lapwings and the rare stone curlew when it was marketed in liquid form. It is now restricted to granules, but these still pose risks to birds.

Thus pesticides do not have to be either new or unusual to pose threats to wildlife. The disastrous decline of British hares has been linked to the use of **paraquat** to clear stubble. Hares are thought to be particularly sensitive to the controversial paraquat and many verbal reports tell of dead hares being found after spraying operations. Dr. Newbold of the Nature Conservancy Council stated in 1978 that 'It's thought safer for mammals but we've often been very mystified where paraquat has gone on meadows. We often get a lot of hare deaths following paraquat spraying, and there is a feeling that it could be far more toxic than is generally appreciated.'

Despite protestations from the manufacturers, ICI, the government's *Approved Products for Farmers and Growers* handbook of 1983 had an additional warning added to its entry on paraquat: 'Paraquat may be harmful to hares, where possible stubble should be sprayed early in the day.' Hares are still being killed, however.

Many other examples exist of animals at risk from pesticides. The Nature Conservancy Council identified **dieldrin** in fish tissues as a likely cause of decline in the otter. Over 80 per cent of otter corpses tested between 1963 and 1973 contained enough dieldrin to reduce their fertility. Its release from industrial processes, has now declined significantly, although it is still a problem in some Scottish rivers. Dieldrin is almost certainly still used illegally as a sheep dip.

> In 1986, more than 25,000 birds died in the Donana National Park in southern Spain after they flew onto neighbouring rice beds which had been sprayed with insecticides. Seven illegal pesticides were used next to one of Europe's largest nature reserves. The dead birds included coots, spoonbills, pintails and mallards.
>
> In 1984, *Farmer's Weekly* reported that dead rabbits were found in a field near Royston, Cambridgeshire, after it had been sprayed with 'Dyfonate' a pesticide containing **fonofos** manufactured by Stauffer Chemical Company UK Ltd. The dead rabbits were found several days after spraying the crop. One theory reported was that fonofos rubbed on to the rabbits' fur, and that they consumed a fatal dose while grooming. The incident was reported to the Ministry of Agriculture.

Poisoning of larger tracts of the countryside

I have dwelt at some length on the side-effects on large birds and mammals, because these are the issues which probably concern the most people. However, I am personally convinced that these are, in fact, a comparatively minor part of the total pesticides and wildlife problem. Deaths of large animals are important, because they often exist in relatively small numbers. Poisoning 100 geese will be much more serious than poisoning 100 mice, voles or sparrows in terms of the overall balance of nature. But deaths of large and obvious creatures are also far more likely to get reported, and acted upon. And, except for the persistent poisons which build up in the food chain, larger animals will usually be more resistant to poisoning than smaller creatures.

The decline in Britain of many small animals, ranging from

mammals to insects, is perfectly obvious to anyone with a few decades experience of the countryside, and is borne out by countless scientific studies. The vast decline in both numbers and species of butterflies is just one example which is particularly noticeable.

There are a number of reasons for this decline. One of the most important is the destruction and degradation of natural habitat through changes in land use such as drainage, felling of ancient woodland and improvement of pasture. This is probably the most significant change to have occurred since the Second World War. Air pollution is another very important factor in the decline of certain types of plants and animals. Hunting has affected numbers of a few species of birds and mammals, including some wildfowl and otter. And, in agricultural areas, pesticides are a very significant factor in this overall decline.

However, although the role of pesticides in killing plants and animals has never been seriously disputed, it is only in the last few years that we have developed methods of showing this experimentally, thus providing hard evidence to back up the vague fears and disquiet felt by many conservationists and country dwellers.

Paradoxically, one of the major driving forces towards greater research into these more subtle pesticide effects has come from a group not usually associated with the conservation lobby at all. The Game Conservancy exists primarily to maintain sufficient stocks of those mammals and birds still hunted for sport or food in Britain. One of the most famous game birds, the grey partridge, is now facing possible extinction, with its population reduced to a third of pre-war levels and still declining. Although habitat destruction (and overshooting) are certain factors in this population collapse, the Game Conservancy scientists believe that pesticides are the single most important cause of falling populations.

The Game Conservancy has monitored population changes over the past 60 years or so. During this period, there has been a steady decline in chick survival rates; that is, the number of chicks reaching six weeks of age. Over the last decade, this has averaged less than 30 per cent and is thought to be a major cause of the overall decline. The partridges are simply not raising enough young to maintain their numbers. Game Conservancy researchers identify starvation as the cause of low survival, and link lack of food to pesticide use.

Pesticides reduce insect numbers in three ways. Herbicides destroy the weeds upon which many insects feed. For example, herbicide used to eradicate grass weeds on small trial plots were found to reduce insects useful to partridges by 43 per cent. Insecticides kill the insects directly, of course, and their use near partridge nesting sites can reduce chick survival as much as 70 per cent. Fungicides can also, as a side-effect, kill the insect food of partridge chicks, both through direct toxicity and by killing the soil fungi on which some insects feed.

Near the Game Conservancy's headquarters at Fordingbridge, Hampshire, scientist Michael Rands has measured the effects of leaving an unsprayed strip (or 'headland') at the edges of the field on Manydown Farm. Results have been fast and dramatic. Headlands left to grow without herbicide had almost three times the wild plant diversity and seven times the number of useful insects. The mean brood size of the grey partridge was significantly greater in areas of unsprayed headlands. In parallel experiments on eight East Anglian farms, mean brood size was almost twice as high as average where unsprayed headlands were left at the margins of fields. And, importantly, there were no significant losses of crop yields as a result.

It was not only the partridges that survived better. Almost three times as many butterflies were recorded on the unsprayed headlands. Species which seemed to benefit particularly from the lack of spraying included the common blue, large and small skipper, orange tip, meadow brown and small heath.

The loss of wild plants

The loss of wild plants is not only a problem for animals which rely on them, directly or indirectly, for food. It also presents a problem for the survival of the plants as well.

In an attempt to reduce the incidence of herbicide drift damage to neighbouring market gardens, the Ministry of Agriculture, Fisheries and Food (MAFF) in Britain issued a list of 'susceptible' crops for which special caution was needed. This includes beet, beans, brassicas, fruit in flower, glasshouse crops, hops, lettuce, ornamentals, peas and vines. Graham Martin, a biologist who has studied spray drift for several years around his home in Evesham, points out that: 'there is nothing special about this list of plants. It covers virtually all broadleaved crops grown in the United Kingdom and all garden plants. If these are susceptible then it would seem reasonable that most if not all natives should be so labelled. It seems clear that the government needs to be persuaded that wild flora is also a "susceptible crop" in need of special consideration by spraying farmers.'

Graham Martin says that farmers in the Vale comment about the decline of once-common shrubs from their hedgerows, without recognising that their own pesticide spray is probably a major contributory factor. In 1981 the ornamental rose was added to MAFF's list of plants particularly susceptible to hormone weedkillers, after a spate of losses from greenhouses totalling thousands of pounds. Yet the wild rose fails to get on to any cautionary list.

The effects of pesticides on aquatic life

Pesticides can also be very toxic to freshwater animals and plants. The most common environmental caution added to information about pesticides is that they are dangerous to fish. In these cases they are very likely to be as dangerous, and usually more dangerous, to aquatic invertebrates as well. For example, the LD50 of **aldrin** for the mosquito *Culex* is as little as 5 parts per billion (ppb). The LD50 of **diazinon** is even less, just 0.9 ppb, for the small freshwater crustacean called *Daphnia*. Larger creatures are also affected; the freshwater shrimp *Gammarus* having an LD50 of 3.8 ppb for the commonly used insecticide **malathion**, while an American toad can be killed by as little as 570 ppb, that is less than one part per million, of the insecticide **endrin**.

> **Diazinon** is an organophosphorus insecticide used to control cabbage root fly, carrot fly, mushroom fly, etc. It appeared in the British '*Approved Products*' handbook, and risks to fish are mentioned. Experiments show its extreme toxicity to aquatic life. For example, the LD50 for caddis flies is around 200 ppb over just

three hours; for mayflies it ranges between 50-134 ppb over 48 hours; and is 48 ppb over 96 hours for one of the molluscs tested. LD50 also changes with the amount of time the diazinon is in the water. The freshwater shrimp has an LD50 of 48 ppb if the diazinon remains in the water for a day, but this falls to 0.5 ppb if it remains for a week.

Deliberate adding of pesticides to freshwater

In the past, pesticides now recognised as being very toxic to fish were deliberately added to freshwater, in Britain and elsewhere, as a means of clearing weeds. Deliberate spraying for weed control, and illegal use as a fish killer, are still widely practised in many Third World countries.

In 1961, scientists with the Norfolk office of the Nature Conservancy (forerunner of the present Nature Conservancy Council) were horrified to learn that the Broads River Authority proposed to clear aquatic weeds from the Norfolk Broads by spraying them with pesticides from a helicopter. The Broads are a nationally important habitat for birds and fish, and the Nature Conservancy feared that the entire ecosystem would be ruined by heavy applications of pesticides. In the event, no insurance company would indemnify the River Authority, so the scheme was abandoned.

Although the mass poisoning of the Broads did not go ahead, other freshwaters were affected. Romney Marsh for example, was sprayed with herbicides for years because it was cheaper than cutting aquatic weeds by hand. As little as 20 years ago the marsh was sprayed by helicopter with both **2,4-D** and **dalapon**, although the former is acknowledged to be dangerous to fish.

Indeed, several herbicides are still cleared for use in freshwater today in Britain, although there are now a special set of guidelines and the local water authority has to be contacted before spraying in or near freshwater. Altogether, seven herbicides are cleared for use on water weeds or shallow water plants and another three for use against grass, bracken and trees on river banks. Of these, five are specifically listed as dangerous to fish (**chlorthiamid, 2,4-D amine, dichlobenil, glyphosphate** and **terbutryn**) 'unless used in accordance with the *Guidelines*. . .'. This means that they are almost certain to be hazardous to other freshwater life.

Accidental additions of pesticides to freshwater

In the rich countries, the main environmental damage from pesticides today comes when the more toxic agrochemicals enter rivers, lakes and streams accidentally. Accidental additions come from seepage from agricultural spraying, direct spray drift, atmospheric deposition (there is often a small amount of pesticide present in rain) and industrial effluents. Accidental spillage causes problems both through occasional large discharges and from the chronic pollution of many watercourses.

In November 1986, a fire in a warehouse owned by the chemical firm Sandoz in Basle, Switzerland, focused public attention on the massive chemical pollution of the River Rhine. The Sandoz warehouse contained about 1,300 tonnes of chemicals, including almost 1,000 tonnes of pesticides and 12 tonnes of organic compounds containing mercury. Hundreds of tonnes of chemicals were washed into the Rhine when fire hoses were played on the factory in an effort to control the fire. In the days that followed, countless thousands of dead fish, including many eels, floated to the surface of the Rhine and virtually all the plankton were destroyed for many miles downstream. The West German environment minister, Walter Wallmann, claimed that the ecology could have been permanently damaged by the accident, and the most optimistic suggestions are that it will be ten years before fish stocks will be anything like normal again. The Dutch measured mercury concentrations of three times normal levels in the 200-kilometre long pollution slick and there were widespread fears for the effects on drinking water extracted from the river.

The accident was significant in its scale, and so it received a good amount of media publicity. But similar accidents take place along the Rhine year in year out. There are literally thousands of small chemical warehouses along the river, and UN officials in Geneva pointed out the impossibility of monitoring all of these. Even while the Sandoz fire was still resulting in severe pollution, several other chemical firms deliberately released unwanted chemicals into the Rhine illegally in the hope that they would not be detected amongst the general levels of pollution. In the fortnight following the fire, five more accidents were reported or discovered polluting air or water along the Rhine, involving such famous companies as Ciba-Geigy (releasing a cloud of phenol); BASF (leaking 1,100 kilograms of the herbicide dichloro-phenoxyacetic acid); Hoechst (850 grams of the solvent chlorbenzol into a tributary of the Rhine); Slefried (a cloud of phosgene gas) and Hoffman La Roche (a toxic liquid, methyl vinylketone). Environmental groups fear that these type of accidents occur all the time, but only get any attention in the aftermath of a major disaster like the Sandoz fire.

In May and June 1984, effluent from a mushroom farm polluted the River Condor near Quernmore in Lancashire. According to the North West Water Authority, **pentachlorphenol**, used as a sterilant in the mushroom plant, had killed life in 200 miles of river. Pentachlorphenol was apparently sold under the trade name 'Santabrite'. This does not appear in the 1986 list of the Pesticide Safety Precaution Scheme and would, therefore, now be illegal. The plant manager, Mr. Ken Drinkwater, was fined £500 by Lancaster magistrates and ordered to pay £150 costs. Since then, canisters of **lindane** have been discovered in a lane near the same plant and some residents believe that further contamination has occurred.

In 1985, a river near Richmond, in Yorkshire, was poisoned when agrochemicals leaked out of a nearby store. About 2,000 brown trout were reported killed, along with several thousand smaller fish. A prosecution apparently took place.

On July 12 1984 a holidaymaker in the Somerset Quantock Hills saw 'approximately 100 empty kilo cartons of Bayleton CF piled up in a stream up to the parapet of a bridge. Two or three fish were seen floating on the water'. The stream ran onto a holiday beach about a mile further along. Bayleton CF contains **triadimefon** and **captafol**. Triadimefon is listed in the *'Approved Products'* book as 'harmful to fish' while captafol does not appear in the book at all and has been banned in West Germany because of its suspected long-term health effects. The boxes carried distinctive warnings about neither using the contents nor disposing of the containers near water courses and about wearing full protective clothing. However, the local authority was reluctant to prosecute and, despite being requested to inform about the result, never apparently followed up the incident.

Thus, there are, indeed, major effects on wildlife from pesticides. These effects are likely to be many times greater in some of the Third World countries, where mass spraying programmes threaten whole populations of animals and where everyday pesticide use is usually far less carefully controlled. Issues such as the mass poisoning of African wildlife in tsetse fly control programmes, and the poisoning of the Amazon, have surfaced in the last few years. For example, in 1985 more than 300 tons of dead fish floated to the surface of two rivers in Brazil's swamp region, and environmentalists claim that herbicides are the cause. These larger scale environmental issues must form a crucial part of any debate about pesticide use in the future.

• What is to be done?

Several immediate steps are needed to protect wildlife in Britain. Some of these are outlined below. In addition, wildlife, like many other 'pesticide victim groups', needs an overall reduction in the amount of pesticides which we use in both the North and South. Britain is currently lagging behind several other European countries in its continued use of pesticides known to be hazardous to wildlife. This unacceptable position must be changed as a matter of great urgency.

- Several pesticides currently in use should be withdrawn altogether because of their effects on wildlife. These include lindane and paraquat amongst others.

- Use of several more, including many organophosphorus pesticides, should be far more restricted than at present.

- The government should put money into schemes to encourage farmers to minimise pesticide use and develop pesticide-free areas on their farms as wildlife refuges.

- Legal protection should be given to nature reserves and Sites of Special Scientific Interest to prevent them being routinely contaminated by pesticide from spray drift from neighbouring land.

- The environmental effects of pesticides should be far more carefully taken into account when devising development aid projects for Third World countries, especially by those United Nations bodies which have been involved in mass spraying programmes of extremely hazardous chemicals in the past.

chapter six

Making pesticides

'Any doctor who is honest in this area will admit the increase in miscarriages is considerable. They are happening in the third and fifth month of pregnancy, and the mothers are crying everywhere. If you go into any one of the hospitals around here – Desio, Giussano, Sevegro, Mariano – and ask where the women who have miscarriages come from, eight out of ten will be Seveso or Meda. This is the first definite symptom we've noticed. It's just not true what the authorities are saying, that the increases in miscarriages has stopped and the ratio returned to normal.'

From *Superpoison* by Tom Margerison, Marjorie Wallace and Dalbert Hallenstein, Macmillan, 1981.

On 10th July 1976, a safety valve jammed in the ICMESA chemical factory at Seveso in northern Italy, a subsidiary of the Swiss transnational corporation Hoffman La Roche. Pressure built up to a critical level and the explosion which followed released 500 kilos of highly toxic vapour into the atmosphere, composed chiefly of **trichlorphenol**, along with **dioxin TCDD**. The latter is the contaminant of Agent Orange, the defoliant 2,4,5-T which has caused such misery in Vietnam.

The local people realised that there must have been a serious accident because they heard the noise of the explosion and could see a reddish brown, foul-smelling cloud enveloping the countryside around the chemical plant. However, they were told nothing. Over the weekend, those living nearby developed a number of symptoms including burning skin and eyes, vomiting, diarrhoea and pains in the kidneys and liver. Four days after the explosion, the company finally released the news that the cloud of gas might be poisonous.

By then, local doctors' surgeries were being crowded with patients suffering from the 'mystery' symptoms of illness. Produce from gardens near the Seveso plant continued to be eaten. However, during the next fortnight, crops began to wither in fields, and birds, rabbits, cats and dogs all died, often in extreme pain. The incidence of human illness continued to increase. Two weeks after the accident, chemical analysis of the cloud was complete, but it was not until four days after that, i.e. 18 days after the explosion, that the first evacuations took place. More evacuations were ordered in the period following this and a number of fairly arbitrary 'zones' of relative danger were delineated. Safety measures remained minimal. Children continued to play in the contaminated dust.

Three weeks after the accident, thirty people were still in hospital. One died on July 29. An acrimonious public row broke out about whether the Catholic Church should sanction abortion in the case of women contaminated by the poison, which is known to be highly teratogenic. Some ignored the laws and did have abortions, but in the autumn following the accident two pregnant women living in the zone nearest to the factory who kept their children gave birth to infants with virtually no brain tissue. Both babies died. Adults continued to experience ill-health.

Incredibly, no proper medical records have been kept of the Seveso disaster. Independent researchers believe that the official figures are consistently pitched far too low. Dr. Alberto Columbi has calculated that birth defects in the whole area were 53 per thousand in 1978, as opposed to official estimates of less than five per thousand for the Lombardy region. Even official figures show an increase in malformed births from a single one in 1976 to 23 in 1977 for births inside the most highly contaminated towns. Many cases of chloracne, the persistent and intense skin acne, developed in the Seveso region.

In February 1977 a 25-year-old district nurse, Maria Bortaccio, developed the symptoms of dioxin poisoning and died in hospital the following October. However, analysis for dioxin poisoning was never carried out on Maria. A local area health official claimed that she had a long history of kidney disease and blamed her death on natural causes. But examination of her hospital records have failed to turn up any evidence that this is true.

Although accurate records have been lost for ever, there is widespread acceptance in Italy that the accident had far greater consequences than ever officially admitted. Dr. Adolph Jann, president of Hoffman La Roche, said of Seveso that 'capitalism means progress, and progress can lead sometimes to some inconvenience'. At Seveso, some people, and some unborn children, were inconvenienced entirely out of their lives.

The accident at Seveso had all the hallmarks of an avoidable disaster, which took place because of corporate indifference and was then clumsily handled after the event. It was already well known that an accident was likely at the chemical plant. The factory had already received numerous warnings about its poor safety record from both local and national government over a 20-year period leading up to July 1976. The delays in evacuation, which took place against the express advice of one senior La Roche safety official, were inexcusable. Lack of official records about the health effects on humans are similarly disgraceful and suggest that some local officials had little interest in seeing the full effects of the disaster publicised. The parent company reacted cynically, as exemplified by Dr. Jann's remarks, quoted above. As one local woman said; 'dioxin follows me around like an evil shadow'.

• Manufacturing hazards

Manufacturing any toxic chemical product is dangerous. Many of the chemicals involved in industrial processes are highly poisonous and, as we have seen they can also pose long-term health hazards to both the workers themselves and their families. The formulation of pesticides is amongst the more hazardous chemical operations and frequently poses high risks to workers in the chemical factories unless particular care is taken. It also threatens people living nearby if there are leaks, explosions or a steady contamination of the environment.

Some pesticide manufacturers take their responsibilities to workers very seriously, and minimise the risks wherever possible. However, in other cases the risks are grossly underestimated or ignored. Such problems do not normally merit much public notice until a major disaster, such as that at Seveso, draws attention to the issue.

The risks which occur during the manufacture of pesticides can be divided roughly into two types, although the distinctions frequently become blurred in practice: risks of very large scale accidents like those at Seveso or Bhopal, and the smaller scale, chronic contamination of workers, the public and the environment. For most people, the chronic problems of pesticide manufacture are actually far more important, even though it is the spectacular disasters which catch media attention.

Mrs. Irene Hogben worked for 18 months as a secretary to Burts and Harvey Ltd. (now Diamond Shamrock Agrochemicals), manufacturers of pesticides based at Belvedere, Kent. Chemicals were supplied to many other firms in Germany, the USA and elsewhere, including 2,4,5-T and 2,4-D. Both are strongly suspected of being teratogens and have been banned from a number of other countries because of this. The chemicals were packed in 50 kilo bags which sometimes broke open releasing a dirty white powder that 'smelled atrocious'. Often men would come into her office covered in white powder, but she was told it was not dangerous. However, she developed a skin problem, in common with many other workers at the plant. Shortly after leaving the firm she became pregnant, and gave birth to a daughter, Kerry, in May 1976. Kerry was born with a massive hole in the heart, a very defective pulmonary artery and with lungs, liver and kidneys too small to function effectively. Kerry needed oxygen to breathe effectively and there was never any chance of her surviving to adulthood. She died in 1983. Mrs. Hogben had a miscarriage in 1978.

Kerry Hogben's sad story illustrates two important sides of the problems of pesticide manufacture. First, factories may have very poor safety standards with respect to pesticides, and secondly, this *laissez faire* attitude can be passed on to workers at the factories, who in turn disregard the possible hazards of chemicals they handle. The general message that pesticides are safe, as promoted by the industry, has the side-effect of reducing the care with which workers handle these products.

Speakers at a conference organised by the Dutch-based Transnational Institute in 1983 illustrate some of the problems of worker safety, and the differences between small and large firms. Workers at Montedison, a small firm based near Florence in Italy, described how the use of small, local subcontractors increased the risks of pesticide manufacture in the area. In order to maintain itself in a competitive market, Montedison put out work to small factories, often with just one assembly line, on an occasional basis. Safety standards were considerably lower in these than in the main plant. There was no union organisation, often no contract of employment, and the firm could enforce compulsory overtime, shorter holidays and hiring and firing in line with short-term needs. This meant that no one was overseeing safety standards, and there is evidence that regulations were frequently ignored in the smaller firms. By using small firms Montedison was allowing lower safety standards to operate, thus probably reducing the price of the pesticides, but increasing worker risk in the process.

Rather different problems were encountered by workers at Rhone Poulenc, a French chemical giant which ranks fifth in the world in pesticide manufacture, and uses 52,000 of its 85,000 employees in this area. Rhone Poulenc has a factory at Dagenham in the UK.

Research workers at the Lyon factory believed that their own safety standards were quite high. But the same was not true in the factory and the need to take precautions was not recognised by many workers. The conference report stated that 'there is also some evidence that disregarding safety regulations somehow enhances the respect of a worker from his workmates', rather like refusing to wear a hard-hat on a building site perhaps. But at the same time, the conference heard that the company's work patterns encouraged this by employing many seasonal workers, made up largely of migrants from southern Europe, who had little idea of the toxic nature of their work.

Thus a factory can expose its workers to risk in two ways. First by having low safety standards itself, and secondly by underplaying the risks involved, or failing to educate workers, so that they do not take sufficient care of themselves.

● The threat of major disasters

The history of chemical manufacture is one which shows that neither the industrialised countries nor the Third World have any grounds for complacency with respect to their record of accidents. Safety standards are often appalling. There have already been numerous accidents and the chances are very high that more will follow in the future. A certain degree of risk is an accepted part of working and living by a chemical factory. The industry gambled on its luck and criticised people who spoke of large scale disaster as alarmist or, as we have seen, subversive. However, the situation was abruptly, and permanently changed in 1984, when the 'world's worst industrial accident' took place at the previously obscure town of Bhophal in India.

A night of terror

Bhopal is a sizeable industrial town in Madhya Pradesh, towards the north of India. It has an estimated population of 800,000 people. In 1969, the chemical company Union Carbide India, a subsidiary of the American transnational with the same name, began building a chemical plant at Bhopal. This was initially involved in just the mixing, dilution and packaging of imported chemicals, i.e. a formulation plant. Later, in 1980, it developed into a full scale manufacturing unit, sprawling over 80 acres towards the edge of town. The chemical plants main product was **carbaryl**, a carbamate pesticide usually sold under the name of **sevin**.

Carbaryl is a dangerous chemical, being an anticholinesterase compound, a suspected carcinogen and teratogen and a skin irritant, and it is also dangerous to fish and wildlife. It may pose special problems for people eating a low protein diet (as is often the case in India, with many people both impoverished and vegetarian). Although carbaryl is acknowledged to be 'moderately hazardous', even by its proponents, one of the intermediary products, **methyl isocyanate**, is known to be instantly toxic and regarded as very dangerous. However, because intermediate substances do not require the same safety clearances as formulated pesticides, neither Union Carbide, nor United Nations agencies monitoring toxic substances, knew very much about its toxicity before the accident at Bhopal.

On the night of December 2, 1984, a massive leak of methyl isocyanate occurred at Union Carbide's Bhopal plant. The precise causes of the accident have still not been fully worked out. Pressure built up inside a tank of methyl isocyanate, probably because of accidental contamination by water. A 60 foot concrete slab, at least six inches thick, cracked open. With a loud hissing sound, a cloud of gas escaped from a tall smokestack into the night air of Bhopal.

Methyl isocyanate is heavier than air. The gas sank from the stack and billowed out amongst the slums of Bhopal, where thousands of people slept in crowded hovels, with no means of sealing off the poisonous fumes. They awoke choking, vomiting and temporarily blind. Those lucky enough to be on upper floors, or in well-sealed buildings, survived relatively unscathed. The others fled. Within minutes, the poisons proved fatal to many and literally thousands died in the street. It is estimated that a quarter of the population of Bhopal, some 200,000 people, inhaled the gas and most of those were injured. Bodies were heaped up into piles, children were separated from their parents, and often whole families were wiped out. By dawn the chaos was so appalling that it was impossible to count the dead and bodies were simply heaped into piles and cremated Hindu fashion.

Official estimates say that between 2-5,000 people died almost immediately. Mr. M. L. Geng, a retired brigadier who runs a factory nearby said: 'my personal estimate is 6-8,000 deaths, bodies were carried off in truck loads, cremated one on top of the other and buried together. There are no records of these people.' Many more victims expired slowly in the days and weeks following, their lungs ruined by the poison. Uncertainty about what gases had escaped, along with clumsy official intervention, hampered the attempts of doctors to treat patients. Many of the 'survivors' have permanently impaired health.

> 'My name is Kallash Pawar. I am a gas victim. I am 25 years old and I work as an auto-rickshaw driver. I was rendered unconscious by the gas, and they put me in a truck with corpses destined for the cremation ground. I awoke with the impact, when I was thrown from the truck on to a heap of dead bodies. My wife died. I have a four-month-old girl, who is not well at all.'
>
> Quoted in *The Bhopal Syndrome* by David Weir, published by the International Organisation of Consumers Unions in Penang, Malaysia.

Bhopal is not only the scene of the world's worse industrial disaster, it is also a particularly unpleasant example of how powerful companies and government officials collude in trying to cover up when inadequate industrial safety precautions lead to catastrophe. We will return to this part of the Bhopal story in the chapter on the politics of pesticides. For now it is sufficient to stress the fact that the accident occurred and the appalling ignorance about the implications of leaking gas once things had started to go wrong. Bhopal will long remain as a potent symbol of all that is worst about pesticide manufacturing.

Although the precise details of death toll and injury will probably never be known, and there is ample room for arguing about numbers and statistics, no one can deny that Bhopal was a disaster of unprecedented proportions. Due to the painstaking detective work carried out by a few journalists and doctors in Italy, the same is probably true of the Seveso explosion. (Indeed, the incident created a whole new EEC Directive.) But the industry argues that these large-scale disasters are nothing more than occasional aberations, which occur because of special sets of circumstances, and are not indicative of what is likely in most cases. They also claim that occasional, regrettable accidents must be separated from any threats to human health from everyday pesticide manufacturing.

Even cursory analysis of the history of the chemical industry shows that this is not the case here. There have been many serious accidents. Those which occur in the Third World frequently get little or no coverage by the media in the North. The health and environmental effects are minimised by the industry wherever they happen. In the table opposite, ten examples of major chemical accidents (not all involving pesticides) are listed as examples, although this is anything but a comprehensive list.

● North and South – who gets the worst deal?

Bhopal also opened up old sores about whether transnational companies take as much care about safety in the Third World countries as they do in the North. Development activists and trades unionists have been saying for years that firms moved into the South partly to get hold of cheap and compliant labour, without a strong union structure, which allows them to cut costs by, amongst other things, ignoring safety standards. This has always been hotly denied by the chemical manufacturers. But after Bhopal they were faced with a quandary. Did they say that conditions in India were just as good (or bad) as in other places? A leak at Union Carbide's West Virginia plant a few weeks after the disaster raised fears of a similar catastrophe occurring there. Or did they say that standards were better in the USA, weakening their case against the massive litigation demanded by the Indian government?

In practice, UC have lost out both ways. It is obvious that safety standards at Bhopal were way below what would be tolerated in Britain or America. But at the same time, the risk of such a large accident happening anywhere are no longer deniable. David Weir quotes Richard Boggs, a management consultant in employee safety and health issues, as saying that 'There is no question that "Bhopal" could happen in the USA. . . . The potential is there and it could happen, maybe today, maybe 50 years from now.'

In November 1985, the Transnationals Information Centre London and the Bhopal Victims Support Group held a conference on the Bhopal tragedy and the European response. Two of the speakers came from the Union Carbide plant in Beziers,

Examples of major chemical and industrial accidents

Place	Date	Details
Oppau, Germany	1921	Explosion at a nitrate manufacturing plant, claiming 561 lives and injuring about 1,500 more.
Texas City, USA	1947	Explosions on two freighters docked in the port and carrying nitrate fertiliser. Most of the port was destroyed and 576 people killed.
Flixborough, UK	1974	Explosion in a railway car carrying cyclohexane, killing 28 people and injuring another 89.
San Carlos de la Rapta, Spain	1978	A truck carrying propylene crashed and burst into flame near a campsite, killing 215 people and injuring several more.
Port Kelang, Malaysia	1980	Explosion of cylinders containing ammonium and oxyacetylene, killing 3 people and injuring about 200 more.
San Juan, Puerto Rico	1981	Two tonnes of chlorine released accidentally from a chemical plant, no immediate deaths but 200 injured.
Montana, Mexico	1981	Derailment of a train carrying chlorine. 29 people killed and well over 1,000 injured by the accident.
Mexico City, Mexico	1984	Explosion of some 80,000 barrels of liquified natural gas, just two weeks before the Bhopal disaster. 452 people killed and 4,248 injured.
West Virginia, USA	1985	Union Carbide pesticide factory suffered a leak of gas, injuring 135 people.
Jabukpar, India	1985	Sodium thiosulphate leaked from tanks in a warehouse, injuring about 100 people.

France, and described a strike which took place in 1977 specifically about safety issues and the use of methyl isocyanate. They argued that Union Carbide knew that the systems were unsafe for years, but continued operating the plant in this fashion for financial reasons. Chemical companies, including Mitsubishi, Du Pont and even Union Carbide themselves, have operated plants without needing to store the deadly methyl isocyante on site, and the plants at Texas, Woodbine and Beziers no longer store MIC. A large part of the blame for the tragedy at Bhopal must lie with the decisions to cut costs, but this occurred because workers were not powerful enough to make their own impact felt on the way the plant operated, and because the government was either not effective enough or not independent enough to tackle Union Carbide before an accident occurred. As long as current practices continue, the chances of more Bhopals occurring in the future remain high.

• *What can you do?*

There is a growing awareness about the need for improving worker safety with respect to pesticides. Government legislation in Britain has sought to tighten the law so that employers have certain obligations with respect to their workforce, starting with the Health and Safety at Work Act in 1974. Several trade unions have full-time safety officers, and employers are obliged to allow recognised unions to appoint safety officers. The manufacturer or supplier of chemicals also has certain legal duties to supply information about chemicals being sold.

However, these regulations are not foolproof by a very long way and substantial worker contamination still continues to occur. There is a need for a substantial overhaul in regulations and greater emphasis on safety in the workplace. The following provides a brief check-list detailing what both employers and employed should take into account with respect to pesticides and other hazardous chemicals.

When dealing with hazardous chemicals, the **employer** should always:

1. Provide a list of all the chemicals being used in a factory, farm or industrial process. This should include chemicals which are not necessarily thought of as 'hazardous', because experience shows that information about chemicals sometimes comes to light a long time after they are in use.

2. Obtain data sheets from the manufacturer. It is important that you check that the chemical composition is listed on the data sheet, because otherwise it is impossible to verify that the risks listed are an accurate summary of hazards. It is always worth cross checking the data sheet if possible.

3. Ensure adequate protection against spillage, etc. and, crucially, sufficient ventilation in the workplace.

4. Provide good quality, and undamaged, protective clothing where necessary. Details of protective clothing requirements are given in some of the publications listed under Books (page 169).

5. Record all human exposures to chemicals amongst your workforce. Complete records of contamination are often the only way of checking back in the case of illness suspected of being caused by exposure to chemicals.

6. Keep to the 'Threshold Limit Value' (TLV) of substances where these are known. TLVs are limits designed to control levels of atmospheric contamination to a safe level. They are defined for about 500 substances, including some pesticides and their constituents. TLVs are sometimes criticised as being inaccurate, but do give some indication of hazard.

7. Reject substances for which there is insufficient data available.

8. Foster an awareness of the hazards of chemicals amongst your workforce.

The **workers** should:

1. Get hold of data sheets from the employer. This is a legal right so if there are problems in getting hold of the sheets, you should contact the factory doctor; complain to the factory inspector; write directly to the manufacturer; contact your union; or refuse to handle the substance.

2. Check the data sheets, these are often inadequate and, if so, you should check back to other sources of information where possible.

3. Use trade union health officers where possible. (Some unions have a good record on providing safety information, others do not.)

4. Always wear protective clothing if stipulated, and take care when handling pesticides.

5. Remember that pesticides can be absorbed in a number of ways, including some or all of: inhalation; swallowing; through the skin; and can cause direct damage to the eyes.

6. Don't smoke or eat while working with poisonous substances. Smoking can result in breathing in fumes which pass through a very high temperature (your cigarette) and this can make them more toxic. Eating or drinking increases the risk of ingesting pesticides.

7. Refuse to work with hazardous substances.

This is a partial listing. For a detailed guide to chemical hazards, try *Chemical Risks* by Maurice Frankel, a Workers' Handbook published by Pluto Press in 1982.

● *What is to be done?*

The need for greater safety in manufacturing pesticides is now obvious. Amongst action needed is:

● Official information about reproduction hazards to be included with all pesticides suspected of causing a risk.

● Threshold Limit Values to be set for all pesticides manufactured in Britain.

● Greater involvement in pesticide safety issues by trade union and employer groups.

chapter seven

Pesticides in the Third World

Pesticide sales are booming world-wide. And the greatest area of growth, if not the greatest area of current sales, is undoubtedly the emerging agricultural countries of the Third World. Coming from a position when many Southern countries used virtually no pesticides at all a few years ago, by 1981 33.5 per cent of global pesticide use was in the Third World and this proportion is still increasing very fast.

The figures are even more significant when the amounts of insecticides used are compared around the world. It is insecticides which, on the whole, cause the most serious problems in terms of human health effects and environmental damage. A 1983 publication from the International Group of National Associations of Manufacturers of Agrochemical Products (GIFAP), based in Brussels, states that insecticides make up only 28 per cent of the USA's total use of plant protectants, and only 20 per cent of Europe's overall consumption. However, insecticides are by far the most important plant protectants used in Africa, Latin America and South-East Asia, making up 63 per cent, 58 per cent and 57 per cent of the totals respectively.

This rapid growth in pesticide sales is not matched by a growth in home-based pesticide manufacturing industry in Third World countries. In 1981, fully 97 per cent of pesticide manufacturers, calculated according to world sales, were still based in the industrialised countries of the North. The Third World relies on imports and, increasingly, on pesticides made in chemical plants in their own countries, but controlled by transnational companies based in the North. Some of the implications of this, and of pesticide use in the Third World, have already been discussed in preceding chapters. In this section, we are going to look at some specific problems relating to pesticide use in the Third World.

• *The Phosvel story*

Phosvel is the exclusive brand name of an organophosphate nerve toxin pesticide called **leptophos**, marketed by the Velsicol Chemical Corporation, which is a subsidiary of the US-based transnational company Northwest Industries. In 1976, worldwide attention was focused on phosvel when the United States Occupational Safety and Health Administration revealed that many of the workers at Velsicol's pesticide plant in Byport, Texas, had developed serious disorders of the central nervous system after involvement with leptophos manufacture. The victims were nicknamed the 'phosvel zombies' because they lost a certain amount of control over their bodily coordination, mental concentration and ability to work. The ensuing media attention focused on the efforts of the workers to sue Velsicol for compensation, and the Texas plant was eventually closed. At that time, Velsicol had never been certified to sell Phosvel in the USA by the Environmental Protection Agency, although the company had received a number of one-year experimental use permits. Its use is now banned in the USA.

In 1980, David Weir and Mark Schapiro of the Center for Investigative Reporting in Berkeley, California, found evidence that Phosvel was still available for sale in a number of Third World countries.

Further investigations uncovered an even more complicated story. Some years previously, USAID, the United States aid and development agency, had paid $4 million to ship 13.9 million pounds of Phosvel to about 50 developing countries in the five years leading up to the revelations of serious health effects in 1976. Large amounts of Phosvel had already been used in the Third World, and news of ill-effects began to trickle back into the North. In Egypt, for example, over 1,000 water

buffalo were killed by Phosvel as early as 1971, along with an unknown number of people.

In 1981, Weir and Schapiro's book *Circle of Poison* laid out clearly for the first time exactly what the US role has been in exporting 'banned' chemicals to poor countries. At that time, pesticides unregistered in the USA (which usually means that they do not match up to safety standards) accounted for over 25 per cent of exports. The vast majority of these are sold to Third World countries where safety standards are lower.

Since then, the debate about pesticide exports from the USA has been growing, although 'banned' pesticides still make up a significant proportion of the exports. Campaigners for greater safety have been enormously helped by the US 'Freedom of Information' legislation, which means that groups can find out exactly where particular products are going. In Europe, and especially in Britain, an obsession with secrecy has helped keep the full details of this end of the pesticide story still firmly under wraps.

● Exports from Europe

The insecticide **dieldrin** is one of the most toxic pesticides known. The World Health Organisation classifies it as 'extremely hazardous'. In 1981, the European Community banned its marketing and use within the member states, although exports are still permitted. Japan banned dieldrin as an ingredient in pesticides 'to prevent environmental contamination' due to its environmental persistence. At least 20 other countries have banned or severely restricted its use, including Canada, Hungary, Turkey and the USSR.

For the Dutch-based transnational Shell, the world's third largest pesticide manufacturer, the failure of the European Community to ban exports of dieldrin as well as use was a financial blessing. Shell is the leading manufacturer of dieldrin and probably makes the majority of global supplies, although some were apparently made at one time in the Velsicol plant of Phosvel notoriety.

> Dieldrin is one of the main 'weapons' in the campaign to eradicate the tsetse fly, organised by the Food and Agriculture Organisation of the United Nations (FAO). Tsetse flies cause about 7,000 cases of sleeping sickness a year in humans, including about 350 fatalities according to WHO, and a vast number of animal deaths, including thousands of cattle.
>
> Tsetse fly health issues are undoubtedly a serious problem. However, there are a number of other problems about using dieldrin to control tsetse fly related illness. First, there is growing evidence that it may not be very effective. Second, it is causing colossal damage to Africa's wildlife. And thirdly, use of dieldrin may itself be resulting in a number of immediate and long-term health problems, especially in view of the extreme persistence which has been the cause of its banning in

many countries. Although the programme of eradicating the tsetse fly has been in operation since 1974, there have been few signs of improvement and FAO officials have been privately voicing doubts about its viability since 1981.

Yet Shell continues to make profits from the export of dieldrin. According to the Dutch group Stichting Mondiaal Alternatief, Shell produced 750 tonnes of dieldrin in 1978 alone.

● Exports from Britain

The British pesticide manufacturers like to make out that exports to Third World countries are a fairly unimportant part of their trade. In the *1984/85 Annual Report and Handbook of the British Agrochemical Association* it was stated that: 'The majority of UK exports by BAA members are to developed countries with full regulatory approval systems. Exports by BAA members to developing countries meet local regulatory approval or those of the international requirements of FAO/WHO.'

However, these bland statements conceal a rapidly growing export market. Between 1979 and 1984 exports from Britain rose from £155.3 million to £480.3 million; i.e. export value has more than tripled in five years. Analysing export figures from HM Customs and Excise Office for 1980, shows that out of a total of £211 million worth of exports of 'disinfectants, insecticides, fungicides, etc.' (i.e. including sheep dips, etc.) some £97 million worth went to the Third World. Nigeria, for example, spent £17 million on British products of this kind. Even some of the smallest states relied heavily on British pesticide exports, with Martinique spending £80,000, St Kitts £33,000 and the tiny British Virgin Island almost £7,000.

The percentage of 'banned' chemicals leaving Britain is said to be much smaller than is the case in the USA, but our official state secrecy makes this impossible to verify legally. If it is true, it is probably more due to the fact that we have not controlled pesticides as carefully ourselves as the Americans, rather than any great difference in export strategy. (This point of view has been put to me by an official in ICI.)

• *Why not sell safer pesticides to the Third World?*

'Dumping' of hazardous chemicals in Third World countries does not just occur out of caprice. However unpleasant and unacceptable the practice may be, there are some hard commercial reasons why pesticide dumping takes place.

Developing new pesticides is a very expensive business. Today's costs for bringing a new agrochemical onto the market average about £20 million. A large proportion of chemicals developed will never be marketed, either because they don't work well enough, or because they fail to pass tests of short-term or long-term safety. Other agrochemicals are withdrawn after being used for some time as we have seen. Many face a limited life in any one country because of the increasing speed with which pests develop resistance to them. Given these mounting costs and uncertainties, it is understandable that chemical companies will fight very hard to maintain sales of a particular product for as long as they can. And if sales are banned in the North, fresh new markets in the South offer a lucrative and easy option for continuing sales in the longer term. At the very least, stocks of banned pesticides can be sold off through foreign markets. At best (from an industrial point of view), powerful transnational companies simply move their manufacturing base over to the South and continue marketing these hazardous materials indefinitely.

• *'A real Mafia-type operation'*

As chemical regulations gradually become tighter in the rich countries, pesticide manufacturing companies have to box clever to keep their products on the market. Exporting banned products becomes more difficult once stocks run low and they are no longer supposed to be manufactured at home. One possible way of keeping hazardous products on the market without attracting too much unwelcome attention is via 'formulation plants' set up in Third World countries.

Formulation plants are not complete pesticide factories, but plants where constituent chemicals are assembled into finished pesticides. Even if a pesticide is banned in its country of origin, the ban seldom extends to its constituent products. These can continue to be manufactured 'at home', making use of existing industrial capacity, and then shipped separately to a formulation plant in a Third World country where no ban on the pesticide exists. Here the finished product is made up and can then be sold within the country, or re-exported to other states where there are fairly lax controls on use of hazardous chemicals.

Described as a 'real mafia-type operation' by Dr. Harold Hubbard of the United Nation's Pan America Health Organisation, formulation plants have sprung up in the last ten years or so. An important reason for this is that they act as a way of maintaining sales of pesticides which are already known to be hazardous.

Most of the biggest formulation plants are owned by transnational chemical companies, including such well known names as Shell, Bayer, Sandoz, Ciba-Geigy, Dow, ICI, Hoechst, Chevron and Union Carbide. Bhopal's ill-fated pesticide factory started out life as a formulation plant, although in this case the pesticide was not banned in the USA.

> The Philippines provide a good example of the weighting in ownership of the means of chemical production. Of the 11 companies owning pesticide formulation plants in the Philippines in 1980, four were from the USA, three from Europe and only four based in the Philippines themselves. A further three American companies sub-contracted other companies to formulate pesticides for them in the Philippines.

The trend towards formulation plants in the South is being copied by many other hazardous industries which now face stricter regulation in the North.

The implications of these developments are all too clear. Not content to dump banned chemicals onto Third World countries, the pesticide manufacturers also want to offload their most hazardous production methods. There are many advantages from their point of view. Labour is cheaper, and is usually without a proper trade union structure to lobby for better standards of industrial safety. Poverty-stricken people are prepared to work in conditions which would not be tolerated in the North. And there are frequently far less stringent safety regulations, or safety standards which are not enforced in practice. The companies can piously rub their hands and lay the blame on unacceptably strict regulations or left-wing agitation as reasons for moving production out of their home country, and have a far more economic proposition in the newly emerging industries of the Third World.

> A plant which manufactures **HCH** in Kenya provides no protection for the workers mixing the chemicals, despite the clear health risks involved. A professor at the University of Nairobi, who visited the plant, described his experiences: 'the workers' eyes are all sunken, and they looked like they had TB. There are regulations against this sort of thing, but there is no manpower for enforcing the regulations. And no one complains. The workers are perfectly happy until one gets sick, and then he's fired.'

As discussed in the chapter on manufacturing, the use of intermediates creates additional problems because their health effects have often not been very carefully studied. The release of methyl isocyanate at Bhopal is now the best known example, but this chemical only made up about 1.5 per cent of pesticide intermediaries worldwide in 1980. Phosphorus compounds, amines, cyanuric chloride, phenol derivatives and aniline derivatives are amongst the commonest intermediates, but there are many more. It was an unwanted by-product, **dioxin**, which caused the health problems at Seveso, as described earlier, and Vietnam, discussed on page 131.

> 'There are manufacturing processes overseas in Third World countries which are detestable. Many places I visit I hate to go into the factory. . . The host governments share responsibility. I'm not exonerating companies from all responsibility. They're perfectly happy to pull out their profits, but the host countries are to blame too. . .'
>
> Gleb Barsky, President of Barsky Sales in Sandy Rock, Connecticut, USA.

The host countries certainly are to blame to some extent, and Gleb Barsky, who describes himself as the 'world's last DDT salesman' has a point when he distinguishes between good and bad companies. But there are other factors which help to keep the Third World in a dangerous position with respect to pesticides. One of these is lack of knowledge of their effects.

• Blowing the whistle on hazardous products

In several chapters of this book, I have described how pesticide users in the South get a worse deal than those in the North. In this chapter, I have shown how the process is intensified by the cynicism of companies which deliberately unload their most hazardous products on to the countries with the least chance of using them safely. But can this process be stopped in the future?

The Farmers Assistance Board (FAB) in the Philippines identifies lack of knowledge about pesticide hazards as a major problem in the country. They describe several examples in their book *Profits from Poison*:

> Nong Enrique Armeglio, his wife and five children all became very ill after eating rice, mangoes and string beans from their farm. According to Nong Enrique: 'I sprayed endrin to our mangoes and string beans in the garden a week before that fateful incident happened. *I really thought it had no residual effect.*' (My emphasis.) A Fertilizer and Pesticide Authority Coordinator reports that: 'Farmers are already used to pesticides. Through constant exposure, the farmers think their bodies will eventually get immune from pesticide hazards. The problem has become a cultural one.'

The FAB argue that this 'cultural problem' is a direct result of the way in which pesticides are advertised and presented. They reproduce adverts which minimise the dangers of using pesticides and show them being applied in totally inappropriate ways. A Shell advert for 'Diagram 5G' shows a smiling farmer broadcasting pesticides by hand without gloves, a mask, or even a long-sleeved shirt. An advertisement by Hoechst shows a young Filipino woman, wearing a sleeveless blouse, standing behind a display of beautiful fruit. Planters Products Inc show a farmer broadcasting lindane without gloves. As the FAB says: 'whether consciously or unconsciously done, such educational materials inform pesticide users that even without the benefit of protective accessories/clothings pesticide is safe to use.'

To some extent, even if such unacceptable practices were banned, development of awareness about pesticide hazards can only be built up over quite a long period of time unless something very spectacular happens in a region to alert people to the facts. The long time scale of chronic effects, like cancer and birth defects, from many pesticides keeps awareness of the problems under wraps for a considerable period. Illiteracy, poor government administration, lack of resources and corruption in both North and South all help to create a situation where accidental or deliberate abuses become much more likely. The general uncertainty of life, and lower life expectancies bring greater fatalism about hazards than is the case with the comfortable people in the North. People facing conditions of desperate poverty and malnutrition have more important things to worry about than long-term health effects, and some Third World leaders have stated this fairly bluntly, including the late Mrs. Gandhi.

However, another very major factor in the lack of regulation in many Third World countries is a parallel lack of knowledge about the status of chemicals being purchased from the North. There are many people in the chemical industry, and amongst their friends in the governments of industrialised countries, who want very much to keep the situation the way it is now. I say this with some confidence, having watched the pesticide proponents fight tooth and nail against several recent attempts to increase the flow of information to the Third World.

One of the longest titles of any report produced by the United Nations Secretariat must be the *Consolidated list of products whose consumption and/or sale have been banned, withdrawn, severely restricted or not approved by governments.* The *Consolidated list* provides information on those industrial, pharmaceutical, agricultural and consumer chemicals which have been severely restricted or banned in countries providing information to the UN researchers. About 60 countries take part. The *Consolidated list* usually includes brief details about why a particular product has been banned and, crucially, gives the trade names of the restricted chemicals. Importantly, it is free, making it accessible to groups and individuals with few resources at their disposal.

All fairly innocuous, you might think. Well no. The decision to prepare the *Consolidated list* was taken at the United Nations General Assembly in December 1982, by a vote of 146:1, with the USA being the sole dissenting voice. Amongst the requirements of the directory were that it 'should contain both generic/chemical and brand names in alphabetical order, as well as the names of all manufacturers. . .' The *Consolidated list* has proved a powerful tool for consumer groups fighting against the use of hazardous chemicals in the Third World, partly because it provides instant access to information about what the rich countries themselves think of their chemical products. This has made it extremely unpopular with some sections of the chemical manufacturing industry. In 1983, *Business International* magazine referred to the book as 'by far the most dangerous step' amongst numerous regulatory activities worrying companies.

In the years following, the USA marshalled a number of other countries to try to 'sanitise' future editions; a specific aim being to remove all the trade names. Pesticides and other products referred to only under complicated chemical names or formulae would be far more difficult to relate to products actually being used in the field, especially for countries where there are no regulations about giving chemical names on packaging. The usefulness (and from the chemical companies' point of view, the threat) of the *Consolidated list* would have been greatly reduced.

Fortunately, a vigorous campaign organised by the Coordinating Committee on Toxics and Drugs, based with the Natural Resources Defense Council in New York, ensured that the directory remains unchanged for the next edition at least. But the lesson is clear. Some powerful people have no wish to see details of regulations in the North become too well known in the South.

The other attempt to guard against chemical companies being able to sell through the buyer's lack of knowledge was a proposed system of 'Prior Informed Consent' (PIC), to be introduced as an international code by the FAO. The principle of PIC is simple. When an exporting country is preparing to sell a pesticide, it would be obliged to provide details of any bans, restrictions or safety precautions legally necessary in the home country. The government of the purchasing country would then have to officially consent to importing the product, after seeing the safety data. This would ensure that details of restrictions reached at least some people in the importing country.

Here, the antipathy of the chemical companies was even stronger and, after vigorous lobbying at the Food and Agricultural Organisation, PIC was rejected in 1985, although at the time this book went to press it was being reviewed. A similar attempt by Oxfam's Public Affairs Unit to introduce PIC regulations into Britain failed during the passage of the 1985 Food and Environment Protection Act, despite support from MPs of all parties. Although the British Agrochemical Association claims that sufficient information will always be given to importing countries, they were adamant that the PIC was unnecessary and lobbied very hard to get the proposed amendments overturned.

There is no logical reason for opposing PIC except a desire to keep buyers in ignorance about dangerous chemicals. I heard the chemical lobby trying to explain away their opposition during the passage of the bill through Parliament on a number of occasions, and the best they could manage was that it would increase bureaucratic problems.

When the pesticide DBCP was banned in the USA, Amvac Corporation continued to sell it in the Third World. As one executive is on record as saying:

> 'The ban on DBCP within the United States was the best thing that could have happened to us. You can't sell it here anymore, but you can still sell it anywhere else.'

However, the Third World does not just suffer from being sold a bad deal with respect to pesticides. In some instances, very large scale misuse of pesticides has been either forced on a country directly, or foisted on them by means of aid policies or some poorly planned 'international project'.

• Agent Orange

By far the most notorious incident involving pesticides, in this case herbicides, being used on people against their will was the mass defoliation of areas of North Vietnam by the USA during the Vietnam War. The story has been told many times, but a brief overview will be given here because of its particular relevance to the pesticides issue.

During the Vietnam War, the Americans became frustrated by the use the North Vietnamese and National Liberation Armies made of forest for cover and hiding. Accordingly, between 1962 and 1971, the US embarked on a mass defoliation programme, spraying a total of some 17 million gallons, or 107 million pounds of herbicides and defoliating a tenth of the country, leaving criss-cross trails of destruction that can still be seen today. Most of the herbicides used were **2,4,5-T** or **2,4-D**. A great deal of the 2,4,5-T was used in a kerosene mixture known as **Agent Orange**. The concentration of **dioxin** in Agent Orange averaged 2 parts per million, but rose to 47 ppm on occasions.

During the defoliation programme, tens, possibly hundreds of thousands of people came into contact with the herbicide. Many Vietnamese were sprayed directly. Numerous American soldiers marched through freshly sprayed areas, or were sprayed by mistake. Others handled the pesticides constantly while they were being applied.

It was known even before 1962 that 2,4,5-T could cause chromosome damage in birds. During the defoliation programme, increasing concern was voiced about the use of the herbicides and their effects in terms of birth deformities and the development of rare cancers like soft tissue carcinomas. South Vietnamese newspapers started to carry stories about a vast rise in birth defects. Massive public indignation eventually led to a suspension of spraying in 1971. In the years since, the evidence for massive health effects has continued to grow. Out of one group of 70 Vietnam veterans studied in the USA, only one had produced a normal first child. As late as 1982, fully a quarter of pregnancies in the main hospital in the Tay Ninh region to the north-west of Saigon were miscarried. A large proportion of the babies which are born there have no eyes or limbs, enlarged heads or no brains.

> 'Nguyen Van Oans was a driver on the Ho Chi Minh trail when it was heavily sprayed. In September 1979 his wife, Le Hu Thin, gave birth to a baby hardly recognisable as a human being. He had anencephalia – no brain. He lived for 28 days and his pitiful body has been preserved for medical science. Two previous pregnancies had resulted in miscarriages. Nguyen Van Oans was sprayed three times with Agent Orange.'
>
> *Portrait of a Poison* by Judith Cook and Chris Kaufman, Pluto Press, 1982.

Eventually, 45,000 Vietnam veterans sued Dow Chemicals, the maker of Agent Orange, and a number of other chemical companies. After years of litigation, a record $180 million was paid by way of compensation. But money cannot buy back dead children, or stop cancers. And no one has paid any money to the people who continue to suffer most, the Vietnamese themselves.

• Aiding and abetting

Normally, the role of the North in the South's use of pesticides is more subtle. I have already described some of these less obvious influences in earlier chapters; the use of USAID to donate the hazardous pesticide Phosvel to 50 Third World countries and the part played by the FAO in the disastrous tsetse fly control programme are two well known examples. Pesticide use has not grown so quickly in the South purely as a matter of chance. There are strong commercial pressures operating in the North as well, and these often go hand in hand with seemingly innocuous influences like our aid programmes and our trade subsidies for Third World countries.

Back in the early 1970s the United States Secretary of State for Agriculture, Earl Butz, stated bluntly that 'food is power'.

In the penultimate chapter of this book, I will look at how pesticides are also 'power', and show some of the ways in which politics and agrochemicals get mixed up in the search for greater profits by the giant chemical companies. Many of these issues strike deep into the heart of the Third World, as described in the latter part of the following chapter.

• *What can be done*

It is certainly not up to us to suggest or impose actions on pesticides on the Third World governments. What we could and should do is to introduce legislation in the North to control the ways in which we currently influence decisions about pesticides in the South. The three most important factors to consider are:

• Introduction of Prior Informed Consent in Britain, as described above, along with support for its implementation into the code of the United Nations Food and Agriculture Organisation at the next possible opportunity.

• Release of information from the Advisory Committee on Pesticides about pesticides which are not passed for use in Britain. This would allow rapid identification of any of those which are subsequently sold abroad, a process which is impossible at present.

• Changes in Britain's aid policy and official attitudes towards the policy of multilateral aid organisations like the European Development Fund and the World Bank, so as to reduce harmful pesticide use and encourage the development of more sustainable agricultural policies.

chapter eight

Are pesticides really useful?

'In the three years between 1974 and 1977, the area of cereals sprayed with aphicides increased 19 times. Between 1979 and 1982, the area of crops treated with insecticides doubled, whilst the area treated with fungicides more than doubled. British Agrochemical Association figures from 1979 to 1982 for the five major crops grown in Britain (cereals, potatoes, sugar beet, oilseed rape and peas) show a 29 per cent increase in the area sprayed with herbicides, a 37 per cent increase for insecticides and a 106 per cent increase for fungicides. Yet the actual cropped area only increased by 4 per cent.'

'Pesticides: An industry out of control' By Chris Rose, from *Green Britain or Industrial Wasteland?* edited by Edward Goldsmith and Nicholas Hildyard, Polity Press, 1986.

The arguments about pesticide hazards and pesticide safety have been see-sawing backwards and forwards for some time. Despite widely divergent views about the risks involved, this debate has normally been carried out in an atmosphere of total acceptance of the necessity of using pesticides, except for 'fringe' groups like organic or biodynamic farmers (who are fringe no more and will be discussed in the chapter on alternatives). The role that agrochemicals play and the current levels of application have seldom been seriously questioned in public, and controversy has focused instead on the safety of individual pesticide formulations and types of spraying systems.

But do we really need to use so many pesticides in the first place? Or any pesticides at all? A growing number of people are starting to believe that pesticide use is, at best, grossly inefficient and sometimes a positive drawback in the struggle to control pests. This chapter examines the efficiency of our pesticide use, and looks at some of the factors which reduce its effectiveness in practice.

I have already described in an earlier chapter how conventional methods of applying pesticide sprays and powders result in the vast majority of pesticides not reaching the pest at all, resulting in waste and pollution. Some of the ways in which this could be improved have also been outlined. But there are other, less easily avoidable, ways in which pesticide sprays are not quite the cure-all that we have often been led to believe.

● *The unwanted side-effects*

The first major inefficiency of pesticides is that many are not at all selective about what they kill, so that an insecticide will destroy many harmless insects along with those which damage the crops being 'protected'. Many fungicides also kill large numbers of insects. In a natural community, numbers of any particular species are kept in balance by several factors, including especially the predatory behaviour of more powerful and aggressive animals. Insects, for instance, are preyed upon by a huge number of mammals, birds, reptiles, amphibians, spiders, centipedes and other insects. Because every predator needs to eat a large number of prey to stay alive, there are always fewer predators than prey. To put it another way, there are a successively smaller number of creatures at each stage up the 'food chain'. Consequently, predators breed more slowly than their prey.

Therefore, many general pesticides kill both pests and their predators. So an insecticide sprayed to kill aphids will also kill the ladybirds which feed on them. But once a field has been cleared of both pests and their predators, it is usually the pests which will breed fastest (either from a few survivors or through immigration from surrounding land). Their predators, on the other hand, build up numbers more gradually and are left with a much reduced population for a longer time. A farmer may easily find that whilst pest numbers drop off dramatically after spraying, they are soon back in even greater numbers than before. At this stage, the usual reaction is to spray again, resulting in another rapid collapse of pests and an equally rapid build up of their numbers again. So the farmer has to spray again. And again. This 'addiction' to pesticides, when the ecology of the farm has been upset, is often called the 'pesticide treadmill'. Once aboard the treadmill, people find it very difficult to get off again.

It is not just insects than can 'rebound' once their natural enemies have been poisoned by pesticides. In Malaysia, the number of owls fell so disastrously after plantation owners used **difenacoum** and **brodifacoum** (described in the chapter on wildlife) that rats multiplied far beyond their original numbers. The pesticides have now been banned and farmers put up owl nesting boxes to encourage their 'natural predators' to return and control the rats again.

Use of these broad action pesticides not only causes rebound of the original pests, but can also result in a more extensive disruption of the ecosystem so that animals not previously considered a problem can themselves develop into destructive pests. At worst, these 'induced' pests can cause more damage than the original pest that was sprayed against in the first place.

Created pests: the spider mite

One of the best documented examples of the 'created pest' is the spider mite, which is now a serious problem in several parts of the world and has been carefully studied in the USA. The tiny, eight-legged, spider mite is related to spiders and scorpions, and feeds by sucking chlorophyll from leaves by means of a specially modified mouthpiece. Mild infestations are not serious to the health of the plant, but with a heavy mite population the foliage turns yellow and fails.

> The spider mite had never been considered a serious problem in the USA until the days of mass forest spraying. Then, for the first time, there were a number of spectacular infestations. Some elementary detective

work soon showed that population explosions of spider mites frequently came in the footsteps of large-
scale spraying operations, often of DDT used against
the spruce budworm. For some reason, the mites are
fairly resistant to DDT (and many other sprays). After
a spraying programme, most of their natural predators, like ladybirds, are dead. And the pesticide
spray, while not killing the mites, irritates them so that
they spread out from their usual dense communities
into all the space vacated by other invertebrates.
Facing little predation, they reduce their usual practice
of spinning protective webs and concentrate all their
efforts on feeding and breeding. The egg production
has been shown sometimes to increase threefold after
spraying. When 885,000 acres of forested land were
sprayed by the United States Forestry Service in 1956,
the worst outbreak of mites on record affected virtually all the treated areas the following year.

Some of these problems can be reduced, if not avoided, by
developing more pest specific chemicals, which destroy a much
narrower range of target species. To some extent, a great deal
has already been achieved in this direction, and far more
accurate chemical killers are available. (In addition, alternatives like biological controls or genetically engineered pest
controls are also more widely available.) Shell, for example,
advertise a pesticide which apparently does not kill the
honeybee.

Unfortunately, there are a number of reasons why development and use of pest specific pesticides does not happen very
often in practice, and farmers continue to rely on broad action
insecticides and herbicides which kill a lot of harmless plants
and animals as well.

One of the main reasons why producers continue to make
broad action pesticides (and thus users rely on these) is that
pesticides are not commodities which have an indefinite life.
Quite the reverse, many have a strictly limited life span due to
the phenomenon of pest resistance.

'From the early 1900s to 1980, 428 species of arthropods are known to have become resistant to one or

more insecticides or acaricides. Over 60 per cent of these species are of agricultural importance, while the remainder are either pests of medical concern or nuisances to people. Among plant pathogens of agricultural crops, over 150 resistant species are known. An estimated 50 herbicide-resistant weed species have now been reported. . .'

Getting Tough: Public policy and the management of pesticide resistance, Michael Dover and Brian Croft, World Resources Institute, Washington DC, November, 1984.

● *New challenges met by changing species*

Many scientists studying the evolution of species now believe that evolutionary change is not invariably a gradual process, as originally suggested by Charles Darwin, but is something which can occur fairly quickly under pressure of a major environmental factor, such as development of a new disease, a change in climate and so on. According to this modification of Darwin's theory, which has been dubbed the theory of 'punctuated equilibria' by Harvard geneticist Stephen Jay Gould, species can stay the same for millions of years then rapidly change when some new pressure kills off all but the minority of 'oddballs' in the population adapted to survive.

If the adaptation is passed on genetically, and the new pressure remains, what was once an unusual feature in the species will soon become the norm and the species will have 'evolved'.

The debate about the causal mechanisms of evolution has been continuing for years, and no end is in sight yet. From our point of view here, it is sufficient to understand that adaptation to a new pressure can take place very rapidly within a population, especially if that population is large (so that the potential for variation is also large) and fast breeding.

Successful pests (which are the ones farmers worry about) fit both the criteria for rapid adaptation; they are numerous and they breed quickly. Some common pests, like aphids or fruit flies, pass through several generations every year and the breeding cycle becomes even faster if food is abundant.

When pests are sprayed with a new pesticide, almost all will be killed if it does its job properly. However, as spraying continues more widely, it is likely to reach a few of the pest species which have a certain natural resistance. These will thus survive the 'eradication programme'. They can then, perhaps, pass on their immunity to their offspring. One or more pockets of resistant pests will soon expand into other areas as the susceptible individuals are killed. Once this happens, the pesticide will become useless for controlling that particular pest.

Pest resistance has been recognised as a problem since soon after the early, heady days of the 'wonder poison' DDT. Indeed, some 229 species of pest had become resistant to DDT by 1980. At one time, resistance was regarded as just one of the risks faced by pesticide manufacturers, and chemical companies hoped that their products remained effective long enough to net them a decent profit. Once resistance developed, they withdrew the formulation and marketed another in its place.

However, several things have made this easy-going strategy less acceptable today. First, the rate with which resistance develops has increased rapidly over the past 20 years. From 1970 to 1980, for example, the number of arthropod species exhibiting resistance almost doubled, while the number of species resisting some particular types of pesticides increased by up to 17 times. Resistance in plant pathogens (like fungi) and in weed species has also grown sharply over this period.

Secondly, some of the more resilient species are developing resistance to more than one chemical, so-called multiple resistance. Thus the total number of cases of arthropod resistance (i.e. number of species multiplied by number of pesticides) is now more than 1,600. Some of these multiple resistant species, including major problem pests like house flies, mosquitos, cotton bollworms, cattle ticks and spider mites, have overcome the toxic effects of just about all the chemicals to which they have been extensively exposed.

The cotton boll weevil disaster

After the Second World War, cotton production in the USA was revolutionised by the use of new **chlorinated hydrocarbon** pesticides such **toxaphene, DDT, benzene hexachloride**, endrin and **dieldrin,** which all destroyed the boll weevil, the primary pest of cotton at that time. They also destroyed the tobacco budworm, which is a secondary pest which only reaches damaging population levels once its natural enemies are destroyed. Soon, the ten to twenty pesticide treatments a year against the boll weevil also destroyed the predators and parasites of the budworm, and the related bollworm, which both became common pests of the cotton for the first time.

Initially, the problem was 'solved' by simply adding a couple of pounds of DDT per acre to the pesticide cocktail specifically for the budworm. When the boll weevil developed resistance to chlorinated hydrocarbons in the 1950s, farmers switched to **methyl parathion** (an organophosphate) for its control instead. However, they still added DDT and toxaphene to their sprays to kill budworms and bollworms, which by that time had become the most serious pests of cotton.

These pests also quickly started developing resistance to these agrochemicals. By 1965, they could not be controlled by any of the chlorinated hydrocarbons (including DDT), or by carbamates like **carbaryl,** or, soon, by methyl parathion. In 1968, despite 15-20 applications of every imaginable combination of pesticide, damage was so severe that many farmers lost virtually all their cotton crop and ploughed in without bothering to harvest. Once this state of resistance was reached, a disastrous decline in cotton production was inevitable. In the Gulf Coast region of Texas, cotton acreage fell from 166,000 acres in 1968 to 55,000 acres in 1975. In the USA, the situation has now been stabilised by far more careful control over the way in which pesticides are used and the introduction of new cotton strains more suitable for pest control. However, in north-western Mexico, cotton acreage fell from 700,000 acres in the 1960s to about 1,000 in 1970, and little cotton is grown in the area today.

• *Cross resistance*

The third risk connected with pesticide resistance is that of 'cross-resistance', where a pest can use the same mechanism to avoid the toxic effects of more than one chemical. This commonly means that a pest can resist several chemically similar pesticides. If several chemical companies produce very similar compounds for use in the same areas or on the same crops they risk a 'domino effect' if resistance to one pesticide develops.

> One group of pesticides causing particular concern are the **synthetic pyrethroids.** These compounds – named after the plant-based insecticide **pyrethrum** which is chemically similar – are toxic to insects in low doses and some (but not all) have low toxicity to mammals. The value of synthetic pyrethroids reached about 1.5 billion dollars in the USA in 1986 and they are estimated to be increasing their sales at an annual rate of about 32 per cent. They are planned to fill a major share in the pesticide market in the future. However, this market takeover is based on the assumption that pyrethroids will remain highly effective for a long time.
>
> Between 1976 and 1980, the number of arthropods becoming resistant to pyrethroids increased from 6 to 17. Out of over 50 species described as promising targets for pyrethroids, about 15 have already become resistant in some part of the world and cross resistance with DDT has been discovered in a few pests. Species resistant to synthetic pyrethroids include bollworm (Australia); cattle tick (Australia); Colorado beetle (USA); German cockroach (USSR); granary weevil (USA); green peach aphid (UK); greenhouse whitefly (UK); house fly (at least five European countries); malaria mosquito (different species in Sudan, Turkey and Pakistan); tobacco budworm (USA); and yellow fever mosquito (Guyana and Thailand). Therefore, many of the most virulent pests have already become resistant to the synthetic pyrethroids which we are putting so much faith into. Yet at least 14 major chemical companies are developing them, or already have products on the market.

No pest control method is particularly likely to avoid the resistance problem. Innovative methods like juvenile hormone mimics, chitin inhibitors, bacterial diseases of insects or natural poisons derived from plants are not immune to the resistance problem.

• *Resistance in the South*

Resistance is not confined to the richer countries. A study in the Cameron Highlands of Malaysia showed that there is complete reliance on insecticides to control the diamond-back moth, *Plutella xylostella*, with applications of mixtures of formulations being applied up to three times a week. Most farmers believe that the diamond-back moth is very tolerant of existing sprays. Doses used sometimes exceed the recommended limits and a quarter of the farmers interviewed admitted to spraying right up to the harvest date.

By 1980 resistance was known or suspected in at least eight insecticides in the region: **DDT, BHC, isobenzan, aminocarb, quinalphos, leptophos, methamidophos** and **fenvalerate.** (In the past, the highly toxic **dieldrin** was also used.) In neighbouring Indonesia, resistance is also suspected in **endrin.** The diamond-back moth is the most important pest of brassicas in the area, and the need for integrated control methods is now accepted.

Resistance becomes a life or death issue when it affects certain species in the Third World. Following the Second World War, the World Health Organisation launched a global campaign to eradicate malaria, by using **DDT** to kill the *Anopheles* mosquito which transmits the disease to people. (I have already described some of the environmental effects of this in the chapter on wildlife.) Initially, success was high and 15 or 20 years after its initiation, WHO officials were talking confidently about worldwide eradication.

However, in the following years, the mosquitoes started developing resistance to various pesticides at an increasing rate. By 1980, 51 out of the 60 malaria-carrying *Anopheles* species were resistant to the three residual insecticides, **DDT, lindane** and **dieldrin,** which were crucial to the eradication programme. At least ten species were also resistant to the organophosphates **malathion** and **fenitrothion**, while far more have multiple resistance to the carbamate **propoxur**. Eighty-four countries now have malaria mosquitoes resistant to at least one of the major pesticides and the incidence of malaria is increasing again, with a doubling of reported cases between 1972 and 1976.

● *The wider implications*

No one is suggesting that all pests will become resistant to all pesticides. But there are a number of real problems created by the increasing role of resistance in the global patterns of pesticide use. First, costs increase sharply, both from increasing crop losses when pesticides fail to work and from the extra price of replacement chemicals needed once the more traditional formulations are 'burned out'. Few estimates of the cost of pesticide resistance have been made, but a World Resources Institute publication of 1984 reported that in the USA additional insecticide application results in costs which annually run to about £130 million from reduced efficiency alone. This figure can probably be tripled if weeds, fungi and the replacement costs of pesticides are included.

On a local level, financial losses can be disastrous and there are now many instances of growers losing a whole crop through resistance. For example, Michigan apple growers lost their apple crops for two consecutive years because apple scab had become resistant to **benomyl**. Benomyl is still extensively used in Britain but the '*Approved Products*' handbook notes that some strains of fungi are resistant. Increased costs have even more acute effects for governments in the Third World who have to pay out for the frequently more expensive replacement pesticides once the old varieties are no longer effective against malaria or other diseases. At the same time, use of pesticides on crops is speeding up this increase in resistance, and many governments find themselves in a vicious circle of pesticide use and replacement which it is difficult to break.

The second, more fundamental, problem is that resistance makes pesticide manufacture a far more risky business. If a pesticide becomes useless before it has generated sufficient revenue, the manufacturing company involved will face serious financial problems. Smaller companies simply do not survive. Although some firms have apparently tried to conceal the existence of resistant pests for as long as possible (thus reducing efficiency of pesticide use even more), once this becomes public knowledge sales inevitably dwindle very quickly.

This has two consequences. First, it is a contributing factor in the increasing concentration of pesticide technology in the hands of a small number of transnational chemical companies. We will return to this point in more detail in the next chapter. And secondly, it encourages the successful companies to concentrate on broad action pesticides. There is little incentive to develop a highly specific agrochemical if the pest concerned can become resistant. Far better to have a general poison which can kill off a whole range of pests, so that if one or more species develop tolerance, sales will not vanish overnight.

This brings us back to where we started on resistance. A lot of the problems for wildlife, problems of rebounding pests, and secondary pests, are caused by the ecological disruption which results from use of broad action pesticides. Yet these pesticides are now becoming common again as a result of the market forces amongst pesticide manufacturers.

This chapter has outlined some of the reasons why pesticides are not as efficient as many people like to believe. They are applied grossly inefficiently. They cause ecological disruption which itself creates fresh pest problems. And pests can become resistant to them surprisingly quickly on occasions.

Pesticide manufacturers will argue that we could not produce as much food without pesticides. Given conventional agricultural systems this may be quite true. But at the very least pesticide use is by no means as efficient as it could be. And more radical scientists now question the whole basis on which conventional farming is based and reject pesticides altogether. In the following chapter, we look at some of the reasons why pesticide use remains high, despite very clear reasons for its reduction. And in the last chapter we look at some of the options, both conventional and unconventional, for the future.

chapter nine

The politics of pesticides

Within a few days of the Bhopal disaster, it became clear that besides (methyl isocyanate), more than 20,000 pounds of broken down products, compounds like hydrogen cyanide, and the even more dangerous cyanogen chlorides had been released from tank No. 610. Nevertheless, a sizeable and influential number of scientists and doctors in Bhopal publicly asserted that no cyanide deaths had taken place. The antidote to cyanide poisoning is sodium thiosulphate (NATS), but thousands of people were given secondary treatment, including large doses of corticosteroids, when they should have been given NATS instead. The medical officers even issued a circular. . . banning the use of NATS.'

Claud Alveres – 'A Walk Through Bhopal', taken from *The Bhopal Syndrome* by David Weir.

Let's look at the Bhopal story again. When future historians assess the disaster, it will probably be as famous (or infamous) for the vicious political battles surrounding its aftermath as it already is for the terrible price paid in human lives.

The political infighting at Bhopal started before the plant was even completed, let alone leaking poisonous gas. Union Carbide planned and built the plant to manufacture **Sevin**, formerly one of their most successful pesticides. However, during the time the plant was being constructed, cheaper (and safer) alternative pesticides became available in India, notably synthetic pyrethroids. Union Carbide's market for Sevin was threatened before they could take advantage of their new manufacturing capacity. Although the company has been accused of trying to head off opposition in India by circulating reports about the health hazards of pyrethroids, sales of Sevin began to fall rapidly. The Bhopal plant barely broke even in its first year of operation and thereafter actually began losing money, operating at half its capacity.

The fact that Union Carbide's pesticide plant was effectively a white elephant probably increased the risk of an accident at Bhopal; there was little corporate interest in the plant within the country and safety margins had been cut back to reduce running costs. In 1982, local journalist Raj Kumar Keswani published a series of exposés in a local paper about various ways in which safety precautions were inadequate at the plant. He also wrote to the Madhya Pradesh States chief minister about the issue, but received no reply. Undeterred, he prepared a further article for *Jansatta*, a popular Hindu periodical, less than six months before the accident. If a massive leak were to occur, he wrote prophetically, 'there will not even be a solitary witness to testify about what took place'.

Concern about safety at the plant was not limited to a maverick journalist. An internal report from Union Carbide officials in the USA, written in 1982, also speaks of 'serious potential for sizeable release of toxic materials'. Workers and trade union officials at the plant itself protested about the dangers they were facing, and about a series of smaller leaks which had occurred before the 1984 disaster.

An analysis of the plant's operational facilities shows that an extraordinary number of safety precautions had simply been ignored. The refrigeration unit which kept methyl isocyanate at low temperatures (and hence much more stable) had been switched off for some time, apparently to save money. The gas scrubber, which would have neutralised the leaking gas, had been turned off for maintenance. (In fact, it would have been incapable of handling a leak of this size in any case.) The flame tower, designed to back up the gas scrubber if any methyl isocyanate escaped the first safety system, was also turned off (and would also have been inadequate to deal with such a large leak). The water curtain, supposed to contain any gas escaping the first two safety nets, was too short to reach the top of the flare tower from which the gas escaped. And the temperature and pressure gauges at the plant were so notoriously unreliable that workers ignored early signs of trouble.

The usual practice was for gas leaks to be 'discovered' when the workers' eyes started to burn and water. This primitive detection system is said to also operate to some extent at Union Carbide's highly sophisticated West Virginia plant in the USA. On the night of December 2, 1984, when workers initially informed a supervisor about the leak, he said they would deal with it after their tea break, an hour later. By then it was too late. An extraordinary catalogue of mismanagement, poor workmanship and lack of corporate responsibility had culminated in the world's worse chemical disaster to date.

In the days following the accident, television news services around the world showed how the disaster's physical reality was almost immediately submerged in squalid financial and political manoeuvring. Sleek, grey-suited lawyers arrived from America to 'represent' the frightened and illiterate slum dwellers who were the main victims of methyl isocyanate poisoning. With claims totalling many millions of dollars, these professional advocates stood to make a fortune from their cut of any damages received.

Elsewhere, Union Carbide and the Madhya Pradesh state government fought to apportion blame. The company faced financial ruin if compensation was paid at the levels being discussed. The national and state governments had to avoid taking the blame for the disaster or they would face bad electoral losses. The president of Union Carbide, Warren Anderson, travelled to India and was temporarily arrested on the Indian government's orders. Union Carbide tried to offer a small relief-fund and was rebuffed because the Indians intended to submit a fuller claim. But despite the posturing, extensive corruption within India meant that many officials had vested interests in clearing the company from the full blame, and minimising the impact of the tragedy.

Union Carbide opened its salvo of defence by blaming (and subsequently attempting to sue) the Madhya Pradesh state government for allowing slums to build up around the plant. It claimed that these were not there when the chemical factory was planned and built, and that they were the reason that so many people were affected by the leak. However, old maps show that at least some of the dwellings were in place before the plant was built. The US parent company claimed that the local subsidiary had failed in its safety precautions. Yet Union Carbide International owned 50.9 per cent of Union Carbide India, despite Indian law limiting foreign ownership of corporations to 40 per cent. In any case, claims that there are differences in safety precautions between North and South is something which the chemical transnationals have always vociferously denied. Union Carbide's international reputation suffered even further when another leak took place at the West Virginia plant on August 12 1985, hospitalising 135 people.

Union Carbide have fought very hard to avoid paying anything like realistic compensation to the victims. In early 1986, they played what may yet be their trump card. At a conference on 'The Chemistry Industry After Bhopal' in the luxurious confines of London's Hyde Park Hotel, J. B. Browning from Union Carbide claimed that the leak had been deliberately started, perhaps by Sikh terrorists, who added water to one of the tanks, initiating the series of reactions which led to the massive leak.

As I wandered around the foyer at lunchtime, the room was buzzing with conversation about the sabotage claim. Comments from delegates, overwhelmingly from within the chemical industry, ranged from 'extraordinary' to 'bullshit'. Given the colossal list of blunders, the previous leaks, and the fact that no one really knew what would happen with the reaction, it is easy to dismiss the claim as pure fantasy. The terrorists would have had to understand a complex reaction better than the plant operators themselves. They would have had to select a chancy and uncertain way of causing an accident rather than simply planting a bomb. The choice of target is totally out of character with terrorist attacks. Yet it is a measure of the power of chemical companies that the claim is receiving serious attention and a detailed case has been filed by Union Carbide in the Indian courts. It does mean that the company can almost certainly stall compensation claims for years to come.

● *Politics and corruption*

It is not the purpose of this book to attempt a detailed investigation into the political manoeuvrings behind global sales of pesticides. The subject is worthy of quite a few books of its own, and some good investigative works are already available. However, in previous chapters I have repeatedly shown that the role of pesticides has too often been determined by their profitability rather than either their safety or usefulness. And, if things go wrong, these immensely powerful companies are able to bring in some heavy guns to avoid problems from public criticism, or national and international legislation, which they would otherwise face. In this chapter, I want to look briefly at some examples of how this power is sometimes abused and at the corruption which occasionally surfaces from beneath the glossy sales literature and smart office buildings which make up the public face of the company.

In 1983, four senior executives of Industrial Bio-Test Industries Inc (IBT), once the world's largest commercial toxicology laboratory, were put on trial in Chicago charged with conspiracy to 'defraud clients and government agencies by writing and distributing false and fraudulent study reports'. The trial, which ended with the imprisonment of the defendants, uncovered an amazing web of corruption and intrigue.

One of IBT's main functions was to test new pesticides to determine their toxicity. This data was used in evaluating their safety, and thus the advisability of registering them in the USA, by the Environmental Protection Agency. IBT staff were proved to have deliberately omitted evidence of hazards from some reports on pesticides and fabricated data (to the extent of replacing dead test rats with live ones). Although they attempted to shred vital evidence once they knew that they were under investigation, sufficient remained available for the charges to stick.

During the course of their work, IBT provided some 800 animal tests for 140 pesticides registered in America. The Environmental Protection Agency claims that three-quarters of those 134 tests are invalid. Some of these chemicals are also used in Britain, although the British government claimed that they had been independently tested.

It still remains a matter of speculation as to why IBT bothered to fiddle their data. The obvious answer, that at least some of the results were falsified on the instructions of one or more companies, remained unproven.

The IBT scandal raised big questions about the honour, and truthfulness, of major chemical companies. It also questioned the impartiality of governments on pesticide testing. Although many countries withdrew the chemicals involved in the IBT tests, at least until they could be re-tested, the British government refused to do this, despite strong pleas from environmental and health groups.

● *Keeping poor chemicals and sprayers on the market*

Waste of chemicals, especially when they are dangerous pollutants, is itself a matter of considerable concern, and is intimately tied up with the political control of some of the methods of application. We have already touched on this in Chapter 8. The British government has, like many others, been content to sit back and watch hazardous chemicals being used extremely wastefully, in ways which are certainly not the most effective in controlling pests. This is in no small way due to the influence that the large chemical companies have built up over the past few decades. There is little political or financial will to improve the efficiency of pesticide use when the most powerful group in agriculture are the chemical producers themselves. The World Resources Institute summed up the position with respect to producing more effective spraying equipment in a report published in 1985:

'The chemical industry's concerns are discovering, developing and marketing pesticides. Pesticide companies control the formulation of active ingredients into marketed products, testing myriad solvents and carriers before releasing the final product. These formulations must meet a host of requirements, such as efficacy and stability in a broad range of environments. Significantly, formulations must usually be "designed to perform in the most readily available equipment used at customary operating conditions". This is understandable: companies seeking to recoup sizeable investments and make a profit don't want to restrict products to ones of new, specialised equipment. Meanwhile, equipment manufacturers want to maximise sales so they design equipment to make use of available chemical formulations. Who should change first, and what's in it for them?'

A Better Mousetrap – Improving pest management for agriculture, Michael J Dover, World Resources Institute, Study 4, September 1985.

Who indeed? At present it is certainly not the chemical multi-nationals, who are ever tightening their grip on the methods of agriculture practised in countries all over the world.

• *Secrecy*

In 1979, the British Royal Commission on Environmental Pollution published their seventh report, on agriculture and pollution, which was highly critical of the secrecy surrounding pesticide hazards in Britain, and their testing by the Advisory Committee on Pesticides. They wrote: 'We think that refusal to release information on grounds of confidentiality tends to become a reflex action, without specific reference to the question of whether commercial interests are truly at stake.'

Since then, an annual report from the Advisory Committee on Pesticides has started to be published as a concession to more openness, and the government promises that summaries of pesticide safety testing data will be released. However, nothing significant has changed. The ACP does not discuss findings on any safety or environmental issues, does not identify pesticides it refuses (so that companies can market them abroad without anyone knowing that they have been refused a licence here) and does not release information on the amounts of particular pesticides manufactured. In addition, the access to raw test data has consistently been refused. Astonishingly, one of the reasons given was that there were often disagreements with the ACP. In other words, it is all right for 'experts' to disagree about whether something is safe or not, but when they finally make a decision no one else is going to have the chance to comment on their results. . .

● The Third World

'Present trends in Philippines agriculture suggest commercialised farming patterned after the agriculture of Western industrialised economies. . . Stress is laid on 'more efficient' farming such as the adoption of high yielding varieties (HYVs), use of machinery, fertilisers and pesticides from the West.

'This relationship did not develop out of necessity but was a deliberate economic imposition on the developing nations by the developed countries. . .

'There was nothing philanthropic in the acts of the Rockefeller, Ford and Kellog Foundations which supported 'scientists' in the research to produce the types of crops needed by the dominant countries of the North. These crops were intended to be produced by Third World nations by exploiting their cheap labour and vast natural resources. . .

'On close scrutiny, the pesticide issue is merely a manifestation, one of the products of a global system of exploitation primarily concerned with profit generation and capital accumulation no matter what the cost – human lives sacrificed or environmental and ecological chaos.'

Profits from Poison, Farmers Advisory Board Inc., Quezon City, the Philippines

The Western chemical companies lay great stress on their role as producers of food for the 'starving millions' in the Third World. Certainly some of their representatives believe they have a vital role in the process of alleviating hunger. But over the last few years, the role of modern agriculture, and of pesticides, in combating malnutrition in the poor nations has come increasingly under attack. The first critical voices were raised by sceptical scientists and politicians in the North. Now, more and more, these criticisms are coming from the South as well. They point to major inconsistencies between what pesticides are claimed to do, and what they achieve in practice.

The much-vaunted 'Green Revolution' of the sixties and seventies was one of the first major development efforts to fall from favour. The Green Revolution, funded by private charities and governments in the North, introduced new, high-yielding crop varieties into the South. Many scientists and administrators became involved for the very best reasons, and believed that they were 'solving' world hunger. However, the new plant varieties relied heavily on fertilisers and pesticides, so that although they would produce higher yields, these were only available to people who could afford the inputs. The richer people produced their higher yields, undercut their poorer neighbours and forced many out of business. The rich got richer and the poor got poorer. There was a concentration of land ownership as a result, because many poor farmers had to sell up. Furthermore, the land was increasingly put over to cash crops, for export, which again benefited the rich minority but did nothing to alleviate general nutrition problems. (It has been shown convincingly many times that there is enough food to feed everyone in the world right now, and that availability of this food is the real problem – political limits rather than resource constraints.) Thus the role of pesticides was to make things worse for those people who could not afford to get onto the bandwagon.

Take India and the Philippines for example, both early recipients of the 'benefits' of the Green Revolution. In both countries food production increased dramatically. However, according to a World Bank staff paper published in 1979, per capita consumption of food grain in India was actually less in 1977 than in 1962. Yet grain production had increased considerably faster than population. In the Philippines, rice production doubled during the 1970s, but the populations' average grain consumption sank during this period to virtually the lowest in Asia, just above that of war-stricken Cambodia.

It is also argued that these increases were not due predominantly to pesticides in the first place, that other factors, like irrigation and mechanisation, played a more substantial role in this process. Indeed, by introducing intensive western-style agriculture into the Third World, some of the more successful traditional methods have been virtually destroyed. The net gain may not be so great as we are usually led to expect.

This is not to say that pesticides have no role at all in the Third World. Pest attacks on crops can cause huge losses and some means of control are important. Pesticides of some kind or another may have a part to play in an integrated pest control strategy. But the emotive use of famine and malnutrition to justify the current pirate tactics of many companies is a stance which is becoming increasingly disreputable in both the North and South.

● *The future*

Whether you view the chemical companies' actions as the immoral machinations of ruthless capitalists or as generally good business in a cut-throat industry which sometimes goes slightly over the top, these actions have succeeded in producing some fairly serious health and environmental effects in the past. An entire system of agriculture has been developed where most farmers are 'hooked' onto a mounting fix of chemicals, and where the rest of us suffer the consequences of their addiction. This situation did not occur by chance alone. It has been brought about by the active efforts of the chemical industry, and frequently by their friends, or representatives, in governments. And breaking this chain, or even introducing some elementary improvements as safeguards, is made more difficult by the practiced resistance from people who do not want to see the slightest erosion of their profits. The rational course of action does not always become the most important in these decisions.

Today, the politics of pesticides is taking a number of new turns. One of the most significant is that the number of companies in the field appears to be shrinking, at least in terms of those capable of developing new pesticide formulations and thus continuing to hold a major place in the world market. Paradoxically, the additional costs of testing new pesticides, along with the increasing problems of developing new formulations, is forcing smaller firms into mergers or bankruptcy. Power is concentrated into the hands of a few major companies. About 24 now control 85 per cent of the total world trade in pesticides.

At the same time, those firms are becoming increasingly involved in other aspects of the food industry, ranging from farming equipment, through food processing and retailing, fast food and, crucially, the ownership of seeds and genetic resources.

I find these developments very frightening. It is easy to assume that because an increasing amount of publicity is focusing on pesticides that things will improve. And so they have in terms of the banning of a few of the most hazardous products. But the future holds the possibility that a few huge corporations, with annual budgets way in excess of all but the largest and most powerful countries, will control virtually all stages of the food cycle. They will produce the seeds to be reliant on their own chemical brands, they will choose the food grain and how it is processed, and they will sell it to us in a form which they decide.

The thought of a few of the most powerful transnational companies controlling the bulk of world food is even more alarming than the current problems of chemical hazards. We have seen that many of these firms are not very bothered about health or environmental impact, nor about the welfare of their workers. The controversy about pesticides will, in the future, inevitably broaden into a more general controversy about who controls food, and exactly how they control it.

• *What is to be done*

There is no simple answer to the abuses of power, or the links between governments and business. Below, some of the more obvious ways of improving some of the specific points made in this chapter are outlined.

- Freedom of information is needed with respect to pesticides in Britain, in line with the recommendations of the Royal Commission on Environmental Pollution.

- Development organisations in the North must use their influence to reduce the grip of Western agribusiness on Third World agriculture. Alternative agricultural methods need far greater promotion through aid programmes, development workers, etc.

- A major international review of the use of pesticides is urgently needed, as part of the preparations for establishing an international organisation to oversee the use of pesticides and other chemicals.

- Full disclosure about commercial links of people on the Advisory Council for Pesticides should be publicly available.

chapter ten

Beyond pesticides: the future of pest control

The golden days of pesticide use are numbered. Although there are many who will argue that the stance taken in this book is too negative, or partial, both professionals and public are far more wary of pesticides than they ever have been in the past. More and more pesticides are being withdrawn around the world, a growing number of governments are wary of automatically accepting the chemical solutions to pest problems, and legislation controlling pesticide use far more carefully is gradually building up in many countries.

This doesn't mean that pesticides are going to disappear overnight, of course, or that our troubles from agrochemicals are almost over. As we have seen, global use is still increasing fast and there will be many more problems before the current types of crop pest control become a thing of the past. However, it does mean that many people, both in conventional agriculture and some of the more radical or alternative movements, are looking beyond the era of chemical farming as we know it.

As might be expected, the views about exactly what should lie beyond pesticides differ very markedly between the various interest groups. They fall very roughly into three options. The first, supported by the chemical industry, other large corporate interests, and many government officials sees the future in the huge advances made in genetic engineering, which they hope will allow the breeding of pest resistant crops, diseases of pests and other 'technical fixes' which will be in a form that can be owned and sold. This is the **biotechnology solution**.

Other interests, including many ecologists, agriculturalists and farmers, see a second path with pesticides continuing to be used at lower levels and in conjunction with many other ecological methods, including use of biological controls, improved crop strains, accurate computer predictions of amounts of pesticides needed and so on. This more sophisticated incorporation of pesticides into pest control is known as **integrated pest management.**

And thirdly, an increasing number of farmers and growers are rejecting the chemical option virtually completely, and developing full-fledged **organic growing** which avoids use of artificial pesticides altogether.

These three options are not mutually exclusive of course. Many governments would pay lip service to supporting a mixture of all three in the future, although the bulk of interest (and most of the profits) are invested in the first option in practice.

These issues are all complicated and require very detailed assessment, beyond the scope of a general book of this type. In this last chapter, these options are examined very briefly, and some of the possibilities and the pitfalls are discussed.

● *Biotechnology*

The modern science of genetics was born when a fairly obscure Austrian monk called Gregor Mendel carried out some ingenious experiments which showed how chacteristics were passed down successive generations of a flowering pea. It took a further, decisive step forward when two researchers at Cambridge University, James Watson and Francis Crick, showed how genes stored and passed on the necessary information for inheritance. Their discovery of the role of deoxyribonucleic acid (DNA) in the genetic code opened up the possibility of understanding and, ultimately, controlling the process of evolution.

The next step took a couple of decades to develop. This was the capacity to isolate chosen genetic material (genes) from one cell and incorporate them into another. When applied to microbes, or cells grown in tissue culture in test tubes, it opened up the possibility of radically changing the characteristics of living materials. For the first time in history, desired genetic material could be taken from one species and added to another, effectively giving the ability to 'create' new species, with physical characteristics fitted to order. This is the set of techniques which have become known as biotechnology or 'genetic engineering'.

The uses that such a technique can be put to are virtually infinite. In relation to human health, the initial priority of genetic engineering was the production of rare human proteins by specially adapted bacteria. Insulin, which must be taken regularly by diabetics, was formerly extracted from the pancreas of cows and pigs in an expensive and tricky operation. Now, a genetically modified bacterium can produce insulin far more cheaply. Other examples still at the experimental stage are the production of interferon and human growth hormones. More widely, genetic engineering is aimed to produce a whole range of services, ranging from the production of growth hormones for cows (which promise increases in milk production from 10-40 per cent), creating new materials from petrol, creating microbes to 'eat up' oil spilt at sea or industrial waste, and so on.

One of the areas where biotechnology is already having the most profound effects is in agriculture, and especially crop production. At present, it takes ten to fifteen years, and a great deal of money, to produce a new crop strain. This process will be able to be enormously speeded up through a mixture of genetic engineering and tissue culture.

Tissue culture involves growing an entire plant from one cell. It allows thousands of identical plants to be produced very quickly. The Dutch multinational Unilever is already growing a million oil plantlets a year through tissue culture, which are then grown to maturity in the company's South-East Asia plantations. When combined with genetic engineering, tissue culture will allow the creation of a whole new strain of plant and, within months, the production of millions of tiny replicas of this. In the case of fast growing plants, today's invention can be this season's main crop.

The possibilities are endless, and there is much speculation about the production of square tomatoes or equal length carrots for ease of packaging, multi-coloured bananas and other science fiction ideas. (Although it must be stressed that these may soon actually exist in the real world.)

There are also more prosaic, and useful possibilities. Principle areas of biotechnological research in crops include: increased rates of growth and yields; nitrogen fixation abilities (thus reducing or eliminating the need for nitrogenous fertilisers); stress tolerance to factors like cold, drought and air pollution; and pest resistance.

● Not such a perfect solution

Despite the possibilities for enormous benefits, there are a number of very serious problems, and potential hazards, with biotechnology. These include physical hazards from the release of genetically altered materials, and political hazards as to how this material will actually be used.

The possibility that scientists will create something they cannot very easily control is one of the earliest and most persistent fears in our industrial society. Experience has shown that the fear all too often has some basis in reality, and many dangerous and destructive substances have been created and released into the environment. Pesticides are a prime example. Biotechnology provides extra possibilities for doing this more often, and of introducing dangerous material capable of reproducing itself. The possible hazards are such that there was a voluntary ban on certain aspects of research for a while but, despite the fact that the questions still remain undebated, the ban has now been lifted. Indeed, we are likely to see an explosion of new 'products' released into the environment, with effects which we have never really considered in much detail at all.

There are also a number of important political (or if you prefer implementational) problems connected with biotechnology. Some of these stem from the fact that the extremely rapid developments in the field over the last few years have centred very much on research funded by, and for, the larger transnational corporations; often the very corporations who are manufacturing pesticides. And their aims may not tally precisely with the wants of the general population. For example, it might be better for a chemical company initially to develop crops which are resistant to pesticides rather than to pests, so that wider use can be made of their chemical products for a few more years until they have generated more profits for the company and can be replaced by crops which do not need pesticides at all.

This is not a fanciful suggestion. Ciba-Geigy recently identified the gene which protects crops against 'Altrazine', one of their most important herbicides. They already sell sorghum seed wrapped in a chemical 'shield' which protects it against 'Dual', another Ciba-Geigy herbicide. And plants produced by tissue culture are said to be more susceptible to pests, so will, at least initially, need more pesticides. The Green Revolution repeats itself on a worldwide scale.

A more serious problem comes from the fact that this ownership of research ability also bring with it the ownership of the improved plant varieties. We have already touched upon the problems of introducing high yield varieties into the Third World. Biotechnology will introduce the same dilemmas into these countries, increasing the potential profitability for those who can afford to get started and further disadvantaging everyone else.

It will also mean that many products traditionally traded from the South to the North will be capable of production in the North. For example, exporting **pyrethrins**, plant-based pesticides, from East African countries currently provides them with much-needed foreign exchange. The same is true of quinine from the Andes, cacao from Africa and many more plant products. These will soon be prepared in the North. It will be argued that this is an inevitable process in any development and that the net gains will benefit everybody. But history shows that when a great deal of money is made in one place, other people (and almost always a greater number of other people) suffer elsewhere.

'There's real genetic imperialism going on here. Let's say that Bolivia has a rare strain of wheat. The peasants have been cultivating it here for 200 years, and it is drought resistant. We can go down there and get that wheat strain – a seed – for nothing. Bring it back to the gene bank in Fort Collins (in the USA), a company will take it out of there, modify it slightly through recombinant DNA, patent it and sell it back to

> the world for millions of dollars' profit, giving nothing
> to the host country.'
>
> Jeremy Rifkin, an American campaigner on genetic
> engineering issues, in an interview in 1986.

Thus biotechnology, for all its promise, is a route fraught with
dangers. It should certainly not be seen as a universal panacea,
as today's optimists and advertising agents tell us. People said
the same about pesticides not so very long ago.

• Integrated pest management

Current pesticide use relies on fairly crude analysis of the pest
problem; a grower has pests and throws down sufficient quantities of pesticide to kill the pests and doesn't mind too much
about wastage; the same pesticides are frequently used
through the season; the time of application is only roughly
calculated; there is little coordination between farms, and so
on.

Integrated pest management attempts to increase the
sophistication of pesticide use. The Office of Technology
Assessment in the USA summed it up well as:

> '... the optimisation of pest control in an economical-
> ly and ecologically sound manner, accomplished by
> the coordinated use of multiple tactics to ensure stable
> crop production and to maintain pest damage below
> the economic injury level while minimising hazards to
> humans, animals, plants and the environment.'

This means just what it says. Using the best control methods
possible whilst minimising risks to living things (including
ourselves) and the environment. This may include using pesticides or it may not. If pesticides are used, they would be
applied in the most sophisticated way possible to avoid waste.
IPM often involves using several methods in tandem, but this is
not a prerequisite if one method will suffice for the whole
problem. And IPM sometimes relies on updated applications
of old methods but, again, this is not always the case.

IPM can include a wide range of pest control methods including:

1. **Biological controls** including both planting patterns to maximise the usefulness of existing predators of pests in the environment (e.g. not using pesticides in a way that also kills the predators), and the deliberate introduction of pest predators into a farming system.

2. **Timing crop planting** to avoid the time of the season when pests will be most active.

3. **Mechanical methods** of killing or avoiding pests, such as use of miniature flame nozzles to kill weeds between crop rows, mechanical strippers for harvest, physical barriers against certain pests, etc.

4. **Removing dead plant material** to avoid build up of pests or their survival over winter.

5. **Rotation of crops** to avoid build up of crop-specific pests in the soil.

6. **Computer technology** to accurately assess precise times of pest infestation, numbers of pests, comparisons with other occurrences, etc, to best plan a control strategy.

7. **Pesticides** applied at the time of maximum usefulness.

And many more. Particular pest infestations will require their own solutions and it is not possible to generalise about exactly what is likely to happen in any one case. A publication on IPM from the World Resources Institute said that IPM 'emphasises systems design rather than knee-jerk reactions to pests'. Rather than just reaching for the spray gun, we should be thinking of other possibilities as well.

In Orissa, India, the gall midge, the brown plant hopper and several other pests were badly damaging the rice crop in the state in the early 1970s. A five-prong attack was designed, including:

- use of early maturing, short-season rice to harvest the crop before the pests' main season;
- monitoring of insect levels to determine when pests passed damage thresholds;

- pest resistant varieties were planted;
- crop residues were ploughed under as soon as the crop was harvested, to prevent stop-over of the pests;
- pesticide was not used at the time of maximum abundance of the pests' parasites and predators.

The cotton boll weevil in the USA, already described in the section on pesticide resistance, has also been tamed through judicious use of the IPM system. The methods used were fairly complex, with ten identifiable components, including:

- planting short season cotton to help eliminate the boll weevil;
- elimination of second or third cuttings to stop the pink bollworm population from building up, and eliminating a potential reservoir for whitefly, which transmit a damaging virus to the cotton;
- irrigation prior to planting in desert areas, where possible;
- planting at such a time as to minimise the number of bollworms which emerge from their overwintering state while a crop is growing;
- monitoring fields for pests and natural enemy populations;
- occasional additions of pesticides at a time to minimise damage to natural predators of the various pests;
- defoliating or dessicating the main crop with chemicals so that all the cotton bolls open together, allowing mechanical harvesting to take place;
- using mechanical stripper for harvest, which kill most of the bollworm larvae;
- applying insecticide during harvesting to catch the remaining adult boll weevils before they migrate to surrounding land and over winter;
- shredding stalks and ploughing under plant remains to deprive pests of overwintering food.

Thus, IPM represents a more considered way of applying pesticides, along with a whole range of other pest control methods, without turning completely away from chemical methods.

• *The organic option*

A growing number of farmers are going one stage further, and abandoning the pesticide option altogether. 'Organic farming' is a system of growing which eliminates all liquid nitrate fertilisers and all but a very few plant-based pesticides. It relies on a knowledge of ecology rather than chemistry to maintain a healthy and sustainable growing system. Plant and animal wastes are recycled through various composting methods to return nutrients to the soil, a healthy soil structure is built up and crops are timed, mixed and rotated in ways which minimise pest attack.

Organic farmers point out that, although some of the methods now used are based upon traditional systems of growing, organic farming is not a return to bygone days. It also relies on the state of the art of knowledge about pest ecology, soil conservation and plant breeding to maximise output.

Elm Farm is a 180-acre holding in the gentle hills of Berkshire near Newbury. It supports about 60 acres of wheat, barley, oats and cereals with the rest in leys for 75 Jersey cows, a flock of around the same number of Poll Dorset ewes and a small beef enterprise. It is unique amongst British organic farms in that it was specifically set up as a research centre, where director Lawrence Woodward and his small team have field trials and laboratory facilities to make detailed comparisons between conventional and organic foodstuffs. Significantly, it is a private charitable trust, and has received no money from central government.

Organic farming has undergone an unprecedented expansion in the last ten years. It has changed from being a strictly minority choice of growing into a system which is attracting a growing number of conventional farmers. Literally hundreds of farms in Britain are now wholly or partly undergoing 'conversion' to organic methods (which itself takes several years to achieve) and many more are showing interest. For the first time, the Ministry of Agriculture has appointed a group to look specifically into the issue. (Britain is lagging behind many other developed countries in this respect, and organic growing is already much further advanced in some European countries.) There are active professional associations for both organic market gardeners and farmers. Wholesale companies move organic food around the country and cannot supply all the demand by any means. An experimental farm is in operation, there are professional journals, interest from within established academic and research bodies, support from MPs and from within the Ministries.

This interest is likely to increase in the future. Current changes in agriculture, food surpluses, the curtailing of grants to support conventional agricultural systems, a general wish for more wildlife areas in the countryside and the increased market for healthy food are all encouraging the development of organic methods. Wholefood shops are now common and several supermarket chains include an organic food section.

What pesticides do you buy if you buy organic?

The most important organic standards scheme in operation in Britain at present is the 'Unified Organic Standards' set by the Organic Standards Committee, coordinated by the Soil Association and with representation from bodies like the Henry Doubleday Research Association, Organic Growers Association, British Organic Farmers and the Farm and Food Society.

The Standards are rigorous. Soil Association inspectors visit any farm before the 'organic symbol of quality' is awarded, and the grower has to abide by a strict set of instructions. Apart from composts or slurry, the only fertilisers allowed are basic slag, ground phosphate rock, dolomite (magnesium limestone), felspar, adularian shale (rock potash), gypsum (calcium sulphate) and ground chalk.

Insect control by biological and ecological methods is encouraged and the only liquid or dust insecticides allowed are pure **pyrethrum, derris** (which is dangerous to fish and thus not allowed near water courses), **garlic, herbal** and **homeopathic** sprays and **quassia.** All are natural preparations from plants. Nicotine was cleared until recently but has been dropped because of its high mammalian toxicity.

Potassium permanganate, sodium silicate and **dispersible sulphur** and **copper fungicides** are permitted. The last two are not ideal because of the problem of their building up in the soil, and it is planned to replace them with a safer alternative as soon as possible. No herbicides, growth inhibitors, sprout inhibitors or combinations of these are permitted at all. Seed dressings are strongly discouraged and mercury seed dressings banned altogether.

Similar strict instructions exist for livestock husbandry. Growth hormone injections and use of antibiotics to stimulate growth are banned and the Standards also include a number of instructions for the humane treatment of animals. For example, debeaking poultry, detailing and teeth cutting of pigs, etc., are banned and housing must be big enough to allow free movement for the animals.

The Standards are not perfect, of course. The Standards Committee admits this and says that it 'is aware these Standards have shortcomings. However, the knowledge and techniques of organic production are rapidly being improved and developed. Consequently we will ensure that our Standards are regularly amended to keep pace with developments.'

Some people argue that, from the point of view of pest management, organic farming is little more than an intensified IPM system. There are obviously similarities in the approach between the two; organic farming also emphasises the systems approach and combines many different factors in farm management. However, there are two essential differences. Organic farming rejects the 'chemical option' altogether, believing that the benefits do not outweigh the risks involved. And, organic farmers are quick to point out, organic growing is about far more than just controlling pests. It involves, at best, a whole new relationship between food production and the rural environment, and a revitalised attitude towards food and health. Some of these issues are now being addressed within bodies like the Soil Association and the links between organic growing and other food issues are being carefully explored.

• *Charter for Agriculture*

In 1984, the Soil Association, Organic Growers Association and British Organic Farmers committed themselves to a 'Charter for Agriculture', which shows that organic growing has wider implications than simply the elimination of artificial chemicals from the farm. The six points in the charter are as follows:

1. Ensure all production and management of farm resources are in harmony rather than in conflict with natural systems.

2. Use and develop technology appropriate to an understanding of biological systems.

3. Rely primarily on renewable energy and diversification to achieve and maintain soil fertility for optimum production.

4. Aim for optimum nutritional value from all staple foods.

5. Encourage decentralised systems for processing, distribution and marketing of farm products.

6. Strive for an equitable relationship between those who work and live on the land, and by maintaining wildlife and its habitats, create a countryside which is aesthetically pleasing to all.

As yet, it is by no means certain which option will be developed in the future. In practice, a mixture of all three is likely. One thing is certain, however, farming has started upon a very considerable process of change, and the traditional role of pesticides is going to be increasingly challenged from all sides in the years ahead.

Books

This book aims to provide a general introduction to the various issues surrounding the use of pesticides. It makes no attempt to be an academic treatise and, accordingly, I have avoided weighing down the text with hundreds of footnotes and individual references. In this appendix I list the main sources for each of the chapters, along with other books and articles which provide accessible background material. If people want to know sources for specific facts or examples mentioned in the book they can write to me, enclosing a stamped addressed envelope, at The Pesticide Trust (see page 176).

There are a vast number of books available giving the chemical farmer's view of pesticide use, and these can be found in any library or bookshop which caters for the agricultural student. In this list, the emphasis follows that of the text, and concentrates on more critical studies.

• Introduction

The book which started off the whole debate about pesticide safety was *Silent Spring* by Rachel Carson, first published in 1962, and still available in Penguin paperbacks. It is now somewhat dated, and a few of her fears proved over pessimistic. However, it remains a powerful and very readable introduction to why everybody should be concerned with the safety of agrochemicals.

Another famous title, of a few years later, is *Pesticides and*

Pollution by Kenneth Mellanby, originally published in Collins New Naturalist series in 1967 and reprinted several times, although now out of print. Worth looking out for in libraries, as it includes an interesting section on the history of DDT and details of early effects on birds.

● Chapter one   **Hazards of pesticides**

There are a large number of handbooks of chemical hazards, many of which include details of pesticides. Few, if any, are even halfway complete, and if you want to follow up on any particular formulation you will have to search through quite a number of different sources to collect information.

Several official books are useful. The *Approved Products for Farmers and Growers* was published annually by HMSO for the Agricultural Chemicals Approvals Scheme (ACAS) until being abandoned in 1985, hopefully to be replaced with something similar in the future. It ignores long-term hazards, but is a useful source of information about short-term effects, harvest intervals, etc., and only 'safer' British chemicals are included.

The *Pesticides Handbook* is published by Blackwells for the British Crop Protection Council and is updated every few years. It gives similar data to the ACAS guide, but includes greater detail.

The *Agrochemicals Handbook* is a large looseleaf compilation from the Royal Society of Chemistry, which has separate data sheets for a large number of pesticides.

Useful guides are also produced by three United Nations bodies, the World Health Organisation, International Labour Organisation and United Nations Secretariat. The last named produced the *Consolidated list of products whose consumption and/or sale have been banned, withdrawn, severely restricted or not approved by governments*, which lists pesticides banned in various countries, along with the reasons. It is available from the UN.

Pesticides by John Watt is the Pesticides Action Network report which provided the basis for the recent Friends of the Earth campaign and is available from them, but at £150 is worth looking out for in a library.

Several books look at individual chemicals, including a number on 2,4,5-T. *Portrait of a Poison* by Judith Cook and Chris Kaufman (Pluto Press, 1982) provides an introduction and a number of case studies, while *The Chemical Scythe* by Alistair Hay (Plenum Press, 1983) is a more detailed account.

● Chapter two   Problems in using pesticides

Conventional methods of pesticide use are outlined clearly in *Pesticide Application Methods* by G. A. Matthews (Longman, 1982).

The Soil Association has published two reports on spray drift problems, including *Pall of Poison* by Vic Thorpe and Nigel Dudley (1984) and *Safety Never Assured* by Nigel Dudley, which focused on aerial spraying (1985).

Friends of the Earth produced a collection of case studies, some of which have been used in this book, which they published as *Pesticides: A first incidents report* by Chris Rose (1985).

The British Crop Protection Council produced their own report on the drift problem, *The Influence of Weather on the Safety and Efficiency of Pesticide Applications* by J. G. Elliott and W. O. Wilson (1983).

Details of the legal requirements of aerial spraying are contained in the *Aerial Application Certificate* available from the Civil Aviation Authority.

For those who do suffer damage, the ADAS publication *Recognition of Herbicide Damage to Crops* by D. J. Eagle (HMSO, 1981) provides colour photographs of a wide range of unintentional damage to crops from herbicides.

Chemical Risk, by Maurice Frankel (Pluto Press, 1982) gives a workers' guide to chemical hazards.

● Chapter three   Pesticides in our food

Contamination of foodstuffs have been well served by two recent accounts. *Gluttons for Punishment* by James Erlichman

(Penguin, 1986) looks at various food contaminants, including pesticides and artificial hormones, providing a readable and balanced introduction. *Pesticide Residues in Food – the case for control* by Pete Snell and Kirsty Nicol of the London Food Commission provides a detailed account of legislative issues, along with a table of several hundred pesticides saying whether they are carcinogenic, teratogenic or mutagenic.

• Chapter four
Pesticides in the home and garden

The Soil Association carried out the first major investigation into garden chemicals during 1986, and the information for the chapter was culled from this research. Further details are available in a pamphlet *How Does Your Garden Grow?* (1986) and a detailed research report *Garden Pesticides* (1986). The latter includes a breakdown of the health and environmental hazards from some 75 pesticides which are used in domestic gardens in Britain.

• Chapter five   Wildlife

In 1985 a major study of pesticides and wildlife in Britain was published: *Pesticides and Nature Conservation: The British experience 1950–1975* by John Sheail (Clarendon Press). This is a valuable source book, but is now rather out of date for today's problems.

Wildlife Link produced a short report on some specific issues, *Pesticides and Wildlife* (1985).

Rachel Carson and Kenneth Mellanby both include details about wildlife in their books, mentioned in the introduction. John Watts' book, mentioned under Hazards of pesticides, also includes a chapter on wildlife.

• Chapter six   Making pesticides

Two excellent investigative books have looked at various aspects of pesticide manufacture.

Superpoison by Tom Margerison, Marjorie Wallace and Dalbert Hallenstein (Macmillan, 1981) is a detailed investigation into the Seveso disaster, produced by members of the *Sunday Times* Insight Team. Although out of print, it is worth looking out for in libraries.

The Bhopal Syndrome, by David Weir (IOCU, 1986, distributed in Britain by Third World Books) looks at Bhopal and at other hazardous pesticide manufacturing, and other chemical industries, in the Third World.

● Chapter seven   Pesticides in the Third World

Several excellent books and pamphlets have focused on problems faced by pesticide users in the Third World.

A Growing Problem by David Bull (Oxfam, 1983) gives a global overview of the problem.

Circle of Poison by David Weir and Mark Schapiro (Institute of Food Policy, 1981) examines the US role in exporting pesticides which have been banned from the home market.

Cleared for Export by Andrew Chetley (Coalition Against Dangerous Exports, 1985) does the same for European exports and includes a chapter on pesticides.

Paying the Price – pesticide subsidies in developing countries by Robert Repetto (World Resources Institute, 1985) looks at the role of subsidies in developing the use of pesticides.

And two classic books, *Food First* by Frances Moore Lappe and Joseph Collins (Ballantine Books, 1978) and *How the Other Half Dies* by Susan George (Penguin, 1976) examine why Western style agriculture does not always work in the Third World.

● Chapter eight   Are pesticides really useful?

Getting Tough: public policy and pesticide resistance (World Resources Institute, Washington DC, 1984). This provides a timely overview of the problems stemming from increasing resistance to pesticides.

● Chapter nine **The politics of pesticides**

Several of the earlier books have touched considerably on the politics of pesticides, including especially those of Lappe and Collins, and George.

The Pesticides Conspiracy by Robert van den Bosch (Doubleday, 1978) is an angry polemic against the industry in the USA.

Since Silent Spring by Frank Graham Jnr (Pan/Ballantine, 1970), although long out of print, is worthing looking out for as it discusses the reactions to Rachel Carson's book and the attempts of some sections of the chemical industry to discredit her. The Campaign for Freedom of Information have produced a number of briefings on pesticide secrecy.

● Chapter ten **Beyond pesticides: the future of pest control**

The subject area is so large that only a few examples from the three main areas are listed as a sampler.

A Better Mousetrap: improving pest management for agriculture by Michael Dover, (World Resources Institute, 1985) looks at improving the efficiency of pesticide use.

And *The Gene Business* by Edward Yoxen (Pan, 1983) casts a critical and informative eye over the whole issue of biotechnology.

New Hope or False Promise? by Henk Hobbelink (ICDA, 1987) focuses on the impact on the Third World.

Several publications of the Soil Association and the Henry Doubleday Research Association describe organic growing. And *Organic Gardening* by Lawrence Hills is a good introduction for the home gardener.

Organisations

A number of British environmental, consumer and worker organisations have put time and effort into campaigning on pesticides, and providing advice to people who have suffered ill-effects. The main ones are:

Friends of the Earth 377 City Road, London EC1 (telephone 01-837 0731)
FOE have a paid pesticides campaigner and an extensive network of local groups, many of which have been involved in lobbying about pesticides. They have published a number of reports about pesticides, including *The First Incidents Report*, and several broadsheets. Work includes Parliamentary lobbying and locally-based campaigns.

London Food Commission PO Box 291, London N5 1DU (telephone 01-633 5782)
Although concentrating on food issues relevant to Londoners, the Commission tackles issues of more general concern. One of these has been the occurrence of pesticide residues in food, and a major report on this subject was published in 1986.

Oxfam 274 Banbury Road, Oxford, OX2 7DZ (telephone 0865-56777)
Oxfam pioneered much of the work on pesticide use in the Third World with David Bull's book *A Growing Problem* in 1982. Since then, their pesticide work in Britain has concen-

trated mainly on the introduction of better export controls for hazardous chemicals.

Pesticide Group c/o Transport and General Workers Union, Transport House, Smith Square, London SW1P 3JB (telephone 01-828 7788)
A group mainly, but not exclusively, of trade union members, interested in the health and safety aspects of pesticides. Organised a major survey of users during 1986 and ran regular meetings.

The Pesticide Trust c/o Earth Resources Research, 258 Pentonville Road, London N1 9JY.
A new group, set up specifically to work on the pesticides issue, and involved in providing information, help and advice to people suffering problems from pesticides, and in lobbying for increased safety measures in British legislation.

The Soil Association 86 Colston Street, Bristol, BS1 5BB (telephone 0272-290661)
The largest organic growing group in Britain, campaigning actively on pesticides for many years, as well as advocating an agriculture which avoids their use. Has published a number of reports on spray drift, aerial spraying, garden pesticides and legislative aspects of agrochemicals. The Association also runs the 'organic symbol' scheme for farmers and market gardeners who wish to sell organic produce.

● *International groups*

In addition, there are a number of international groups involved in the pesticides issue. The most important of these are:

The Pesticides Action Network (PAN) an international organisation with offices in five continents:

Pan Europe, 22 rue des Bollandistes, 1040 Brussels, Belgium. PAN groups function independently, and each regional office includes many member organisations within that area, making PAN by far the largest pesticide group in the world in terms of total associated membership. PAN also mounts international campaigns, such as the 'Dirty Dozen', which focused attention on 12 particularly toxic chemical products.

Friends of the Earth International PO Box 17 170, 1001 JD, Amsterdam, The Netherlands (telephone 31-20-340 252) Coordinates campaigns between the 40 or so national FOE groups, including organisations in countries as diverse as Australia, Hong Kong and Papua New Guinea. Some national groups, e.g. Malaysia, have been very active on the pesticides issue.

International Coalition for Development Action (ICDA) 22 rue des Bollandistes, 1040 Brussels, Belgium A group with extensive contacts in the Third World and a history of involvement in pesticide issues.

Transnationals Institute Paulos Pottertsr. 20, Amsterdam, 1071 DA, Netherlands (telephone 31-20-726 6080)

Glossary

Acaricide
Pesticide used against spiders and mites

ACP
Advisory Committee on Pesticides, government committee of independent experts who determine if pesticides are safe enough to be allowed for use.

ACAS
Agricultural Chemical Approval Scheme, now defunct second scheme for assessing health risks and suitability of pesticides in Britain, which produced the annual "Approved Products" guide.

ADAS
The Agricultural Development and Advisory Service, part of MAFF charged with giving advice to farmers and growers. Being successively cut back and privatised.

Allergen
Substance likely to cause an allergic reaction in susceptible people.

Anti-cholinesterase
Substances which inhibit the enzyme cholinesterase in susceptible people, causing muscle spasms and acute poisoning.

BAA
British Agrochemicals Association, trade group of pesticide manufacturers.

BASIS
British Agrochemicals Standards Approval Scheme.

BCPC
British Crop Protection Council, industry body which organises annual conferences, publishes annuals, occasional research reports etc.

CAA
Civil Aviation Authority, part of the Department of Transport overseeing civil aviation, including aerial application of pesticides.

Carcinogen
A substance capable of stimulating the formation of cancer.

CDA
Controlled Droplet Application, a relatively recent innovation allowing pesticide to be applied more accurately and with less waste, by controlling droplet size to the optimal for particular conditions. One common method is to use a spinning disc to form the droplets.

Codex Alimentarius Commission
A standing body sponsored jointly by FAO and WHO to set an international list of MRLs.

Electrostatic spraying
Method of applying pesticide spray by giving all the droplets weak electric charge so that they are attracted to plants, allowing greater accuracy and less waste.

EPA
Environmental Protection Agency, US body charged with the whole range of ecological issues, including pesticide safety.

FAO
Food and Agriculture Organisation, United Nations body based in Rome overseeing the UN's work in agriculture and forestry in the Third World. Concerned with safety of pesticides.

FDA
Food and Drug Administration of the USA.

Food and Environment Protection Act
British Act of Parliament in 1985 which was the centre of enormous debate about the future of pesticides.

Fungicide
Pesticide used against fungi and moulds.

Herbicide
Pesticide used against weeds and grasses.

HMSO
Her Majesty's Stationery Office, the British government's official publishing house, bringing out all official publications, also books from various official departments, etc.

HSE
Health and Safety Executive, official body charged with health and safety issues.

IARC
International Agency for Research into Cancer

ILO
International Labour Organisation, United Nations body interested in working people and employment, publish a detailed manual on safety of commercial chemicals, including pesticides.

Insecticide
Pesticide used against insects (and often other arthropods).

IPM
Integrated Pest Management, modern system of pest control which minimises pesticide input.

IRPTC
International Register of Potentially Toxic Chemicals.

LD50
Dose of chemical needed to kill 50 per cent of test animals, usually measured in mg/Kg body weight for animals like rats or rabbits.

MAFF
Ministry of Agriculture, Fisheries and Food, the British government's ministry dealing with agricultural issues.

Molluscicide
Pesticide used against slugs and snails.

MRLs
Maximum Residue Limits on pesticides.

Mutagen
Substance capable of causing genetic damage and thus causing possible problems in future generations.

NCC
Nature Conservancy Council, official body charged with overseeing nature conservation in Britain.

Neurotoxic
Toxic to the nervous system.

NFU
National Farmers Union.

NIOSH
National Institute of Occupational Safety and Health in the USA.

PSPS
Pesticides Safety Precaution Scheme, official scheme for clearance of pesticides, long voluntary and now a legal requirement before pesticides are allowed to be used.

RCEP
Royal Commission on Environmental Pollution.

Rodenticides
Pesticide used against mice and rats.

SSC
Scientific Sub-Committee of the ACP, advising on pesticide safety.

SSSI
Site of Special Scientific Interest, area designated by the Nature Conservancy Council (NCC) and being of particular conservation interest, with controls over how it can be used.

Systemic fungicide
Can prevent development of fungus away from the point of application.

Teratogen
Substance capable of causing birth defects.

Translocated herbicide
Transported within the plant and can affect parts remote from the point of application.

ULV
Ultra Low Volume, method of applying pesticide using minimum volume of dilutant liquid to carry the agrochemical and usually using very small droplets.

UNEP
United Nations Environment Programme.

WHO
World Health Organisation, United Nations organisation charged with working on global health issues, including safety of pesticides.

THE PESTICIDE TRUST

The Pesticide Trust has been formed by a coalition of environmental groups, health organisations, development groups and trade unions. It will act as a central organisation to provide information about pesticide hazards and safety issues, to carry out and co-ordinate research, liaise with similar organisations in other countries and lobby for better safety standards for pesticides in Britain.

The Trust will publish regular bulletins, papers and larger publications and will have both individual members and membership by organisation.

For more information about the Pesticide Trust, please write to:

> The Pesticides Trust,
> c/o 258 Pentonville Road,
> London N1 9JY

Appendix

APPENDIX   Checklist of Pesticide Hazards

	Classification	Poison	Carcinogen	Terato/Mutagen	Anti-Cholinester	Irritating	Harvest interval	Danger to wildlife	Danger to fish	Danger to bees	Danger to plants	Persistence	Resistance	Banned/restricted	Garden use
Aldicarb	I	•		•	•		•	•	•	•		•		•	
Aldrin	I	•	•	•					•		•	•		•	
Alloxydim-Sodium	H					•	•					•			•
2-Aminobutane	F	•				•				•					•
Aminotriazole	H	•	•	•		•			•		•			•	•
Amitraz	A	•					•		•						•
Ammonium sulphamate	H					•				•					•
Asulam	H	•							•	•				•	•
Atrazine	H	•		•		•		•	•	•	•	•	•		•
Azinphos-Methyl	I	•	•	•	•		•	•	•	•	•	•	•		•
Aziprotryne	H						•		•						
Benazolin	H								•		•				
Bendiocarb	I				•			•	•	•					
Benodanil	F						•		•	•	•	•			
Benomyl	F	•	•	•		•						•	•	•	•
Bentazone	H					•			•		•				
Benzoylprop-Ethyl	H								•		•				
Bifenox	H								•		•				
Binapacryl	F			•			•	•	•			•	•		
Bioresmethrin	I									•					•
Bromacil	H					•						•		•	
Bromophos	I	•					•	•		•		•			
Bromoxynil	H								•		•				
Bupirimate	F	•				•	•		•					•	•

184

	Classification	Poison	Carcinogen	Terato/Mutagen	Anti-Cholinester	Irritating	Harvest interval	Danger to wildlife	Danger to fish	Danger to bees	Danger to plants	Persistence	Resistance	Banned/restricted	Garden use	Restrictions
ptafol	F		•	•		•						•		•		
ptan	F	•	•	•		•			•	•	•	•		•	•	
rbaryl	I	•	•	•	•	•	•		•	•		•		•	•	•
rbendazim	F	•	•	•		•						•				
rbetamide	H						•	•								
bofuran	I	•	•	•	•				•	•						•
rbon tetrachloride			•	•										•		
rbophenothion	I	•			•				•	•		•		•		•
rbosulfan					•				•	•				•		•
arboxin)	only available for use with organomercury compounds															
lorobromuron						•			•			•	•			
ordane	I	•	•	•		•		•	•	•	•	•	•	•	•	•
lorfenvinphos	I	•		•	•			•	•	•		•				•
loridazon	H								•			•	•			
ormequat	GR															
hloroethyl-Phosphonic Acid	GR					•			•			•				
lorothalonil	F					•	•		•			•		•		
loroxuron	H					•										
orprophan	H															
orpyrifos	I				•	•			•	•	•	•	•			
orpyrifos-Methyl	I			•	•				•							
orosulfuron	H			•	•				•			•				
orthal-Dimethyl	H					•						•				
orthiamid	H						•		•			•				•
ortoluron	H					•						•	•			
pyralid	H					•	•					•				
pper	F							•	•			•	•			
anazine	H		•						•			•				
cloate	H					•			•			•	•			

185

	Classification	Poison	Carcinogen	Terato/Mutagen	Anti-Cholinester	Irritating	Harvest interval	Danger to wildlife	Danger to fish	Danger to bees	Danger to plants	Persistence	Resistance	Banned/restricted	Garden use
Cyhexatin	A					•			•			•	•		
Cymoxanil	F					•	•		•						
Cypermethrin	I					•			•	•					
2,4-D	H		•	•		•			•		•			•	•
Dalapon	H					•						•			•
Dazomet	SS					•						•			
2,4-DB	H						•		•			•			
Deltamethrin	I	•				•			•	•					
Derneton S Methyl	I	•		•	•		•	•	•	•	•	•		•	•
Derris	I					•			•	•		•			•
-Desmediphan	H					•									
Desmetryn	H						•					•			
Diazinon	I			•	•	•	•	•	•	•	•	•	•		•
Dicamba	H		•			•			•		•				•
Dichlobenil	H					•			•			•	•		•
Dichlofluanid	F	•	•				•		•			•			•
Dichlorphen	H	•				•			•		•				•
Dichlorpropene	N					•						•			
Dichlorprop	H	•				•			•		•				•
Dichlorvos	I	•	•	•	•	•	•	•	•	•		•			
Diclofop-Methyl	H						•	•	•						
Dicoful	A	•	•				•	•	•	•	•	•	•		•
Dicloran	F											•			
Dieldrin	I	•	•	•		•				•	•			•	
Difenzoquat	H														
Diflubenzuron	I						•								
Dimethirimol	F											•			
Dimethoate	I	•		•	•	•	•	•	•	•		•	•	•	•
Dinobutan													•		

	Classification	Poison	Carcinogen	Terato/Mutagen	Anti-Cholinester	Irritating	Harvest interval	Danger to wildlife	Danger to fish	Danger to bees	Danger to plants	Persistence	Resistance	Banned/restricted	Garden use	Restrictions
Dinocap	F	•		•		•	•		•	•	•	•	•		•	
Dinoseb	H	•		•		•	•	•	•	•	•					•
Diphenamid	H								•		•					
Diquat	H	•		•		•	•	•	•				•		•	
Disulfoton	I	•		•	•	•	•	•	•	•	•	•		•		•
Dithianon	F					•						•				
Diuron	H					•			•			•	•			
DNOC	I	•				•		•	•	•	•	•				•
Dodemorph	F					•			•			•	•			
Dodine	F			•		•						•				
2,4 D-P	see Dichlorprop															
Drazoxolon	F	•				•	•	•	•			•				•
Endosulfan	I	•				•	•	•	•	•	•	•			•	•
Endrin		•	•	•	•	•			•	•		•			•	•
EPTC	H					•			•			•				
Ethirimol	F	•				•						•	•	•		
Ethofumesate	H						•		•			•				
Ethoxyquin	AO															
Etridiazole	F			•		•	•					•	•			
Fenarimol	F					•			•						•	
Fenitrothion	I	•		•	•	•	•	•	•	•		•			•	•
Fenoprop	H					•						•			•	
Fenpropimorph	F					•	•		•							
Fentin	F	•						•	•			•				•
Ferrous sulphate	H			•								•			•	
Flamprop-M-Isopropyl	H					•			•							
Flamprop-Methyl	H					•			•							
Fluazifop-P-Butyl	H					•	•					•				
Folpet	F		•									•	•		•	•

	Classification	Poison	Carcinogen	Terato/Mutagen	Anti-Cholinester	Irritating	Harvest interval	Danger to wildlife	Danger to fish	Danger to bees	Danger to plants	Persistence	Resistance	Banned/restricted	Garden use	Restrictions
Fonofos	I	●		●			●	●	●	●		●		●		●
Formothion	I	●		●		●			●	●		●			●	
Fosamine ammonium	H													●		●
Gamma HCH	I	●	●	●			●	●	●	●	●	●		●	●	
Glyphosphate	H					●	●	●	●	●		●	●		●	
Guazatine	F	●							●							
Heptenephos	I	●			●		●	●	●	●						
Hexazinone	H					●						●				
Imazalil	F						●		●	●	●					
Iodofenphos	I				●		●		●							
Ioxynil	H			●		●			●		●				●	●
Iprodione	F						●		●		●					
Isoproturon	H					●					●	●				
Lenacil	H					●			●		●	●				
Lindane	–see gamma HCH															
Linuron	H					●			●		●	●				
Malathion	I	●		●	●	●	●		●	●	●	●	●		●	
Maleic hydrazide	H		●	●					●					●	●	
Mancozeb	F	●		●		●	●					●			●	
Maneb	F		●	●		●	●		●			●		●	●	
Mn. Zn Ethylene Bisdithiocarbamate	F					●	●									
MCPA	H	●	●	●		●			●	●	●				●	
MCPB	H								●		●					
Mecoprop	H					●					●				●	
Mefluidide	GR					●		●			●					
Mephosfolan	I	●			●		●	●	●	●		●		●		●
Mepiquat chloride	GR					●			●		●	●		●		
Mercuric oxide	F	●							●			●		●	●	●
Mercurous chloride	F	●						●	●	●		●		●	●	●

	Classification	Poison	Carcinogen	Terato/Mutagen	Anti-Cholinester	Irritating	Harvest Interval	Danger to wildlife	Danger to fish	Danger to bees	Danger to plants	Persistence	Resistance	Banned/restricted	Garden use	Restrictions
Mercury	F	•				•		•	•	•	•			•		•
Metalaxyl	F					•	•		•					•		
Metaldehyde	M	•				•		•	•	•					•	
Metamitron	H											•	•			
Methabenzthiazuron	H								•			•				
Metham-Sodium	SS					•						•				
Methazole	H					•										
Methiocarb	H	•			•	•		•	•	•		•			•	
Methomyl	I	•			•	•		•	•	•	•	•		•	•	•
Metoxuron	H											•				
Metribuzin	H											•	•			
Mevinphos	I	•			•	•		•	•	•	•	•			•	•
Monolinuron	H					•		•	•			•	•			
Nabam	F					•	•									
1 Naphthyl-Acetic acid	GR											•				
Napropamide	H					•			•			•				
Nicotine	I	•	•	•				•	•	•	•	•			•	•
Ofurace	F					•	•		•							
Omethoate	I	•		•	•	•		•	•	•		•				•
Oxamyl	I	•			•	•		•	•	•		•				•
Oxycarboxin	F											•	•	•		
Oxydemeton-Methyl	I	•			•	•		•	•	•		•			•	•
Paraquat	H	•		•		•			•			•		•	•	•
Parathion												•				
Pendimethalin	H					•			•			•				
Pentachlorphenol	F	•	•	•		•										
Permethrin	I	•	•			•			•	•		•			•	
Phenmedipham	H								•			•				
Phenylmercury acetate	F												•			

	Classification	Poison	Carcinogen	Terato/Mutagen	Anti-Cholinester	Irritating	Harvest interval	Danger to wildlife	Danger to fish	Danger to bees	Danger to plants	Persistence	Resistance	Banned/restricted	Garden use	Restrictions
Phorate	I	●			●	●	●	●	●	●	●	●			●	●
Phosalone	I				●		●	●	●			●	●			
Picloram	H		●			●			●			●	●		●	
Piperonyl butoxide	C		●												●	
Pirimicarb	I	●			●		●	●		●					●	
Pirmiphos-Methyl	I	●		●	●	●	●		●	●			●		●	
Prochloraz	F					●			●		●					
Prometryn	H			●		●						●	●			
Propachlor	H			●		●						●				
Propham	H		●			●			●			●	●			
Propinconazole	F	●				●	●		●						●	
Propoxur		●										●				
Propyzamide	H					●	●					●	●			
Pyrazophos	F			●	●	●	●	●	●	●	●	●	●			
Pyrethrum	I					●						●	●		●	
Pyridate	H					●										
Quinalphos	I	●			●	●	●	●	●	●						●
Quinomethionate	A/F					●						●	●			
Quintozene	F					●						●	●	●	●	
Resmethrin	F	●							●	●		●			●	
Simazine	H	●		●		●			●	●	●	●	●	●	●	
Sodium chlorate	H	●		●		●						●	●		●	
Sodium monochloracetate	H	●				●				●	●	●	●		●	
Sulphur	F					●						●	●		●	
2,4,5-T	H	●	●	●				●	●			●		●	●	
Tar oil	I	●	●			●			●			●			●	●
2,3,6-TBA	H					●			●			●				
TCA	H					●						●	●			
Tebutam	H					●			●			●				